Policing and Victims

❖

LAURA J. MORIARTY

Editor

M.L. DANTZKER

Series Editor

Prentice Hall

Upper Saddle River, New Jersey 07458

Library of Congress Cataloging-in-Publication Data
Policing and victims / Laura J. Moriarty, editor.
    p. cm. — (Prentice Hall's policing and . . . series)
  Includes bibliographical references (p.).
  ISBN 0-13-017920-5 (alk. paper)
    1. Victims of crimes—United States. 2. Criminal justice, Administration of—United States. I.
Moriarty, Laura J. II. Series.
  HV6250.3U5 P647 2002
  362.88′0973—dc21

00-064326

Publisher: Jeff Johnston
Senior Acquisitions Editor: Kim Davies
Production Editor: Janet Kiefer, Carlisle Publishers Services
Production Liaison: Barbara Marttine Cappuccio
Director of Manufacturing and Production: Bruce Johnson
Managing Editor: Mary Carnis
Manufacturing Buyer: Cathleen Petersen
Art Director: Marianne Frasco
Cover Design Coordinator: Miguel Ortiz
Cover Design: Wanda España
Cover Photo: ©Tracey L. Williams/Courtesy of Somerset County Police Academy,
  North Branch, NJ (Dr. Richard Celeste, Deputy Chief, Executive Director); and
  Jersey Battered Women's Service, Fair Lawn, NJ (Phyllis Canzona, MA, MSW,
  Coordinator of Education and Training).
Marketing Manager: Ramona Sherman
Editorial Assistant: Sarah Holle
Interior Design: Monica Kompter
Composition: Carlisle Communications, Ltd.
Printing and Binding: R. R. Donnelley & Sons

Prentice-Hall International (UK) Limited, *London*
Prentice-Hall of Australia Pty. Limited, *Sydney*
Prentice-Hall Canada Inc., *Toronto*
Prentice-Hall Hispanoamericana, S.A., *Mexico*
Prentice-Hall of India Private Limited, *New Delhi*
Prentice-Hall of Japan, Inc., *Tokyo*
Prentice-Hall Singapore Pte. Ltd.
Editora Prentice-Hall do Brasil, Ltda., *Rio de Janeiro*

10 9 8 7 6 5 4 3 2 1
ISBN 0-13-017920-5

Dedication

To my godmother
Helen Wagner

"How natural it is that I should feel as I do about you, for you have a very special place in my heart."

Philippians 1:7

Table of Contents

❖

PART III
RESOURCE ISSUES

Preface

Police officers are the first representatives of the criminal justice system victims encounter. In many instances, these first encounters result in conflict. For the most part, when victims contact the police, they expect immediate results. Police officers, conversely, expect victims to provide accurate reporting of the events that led to the call for service. These two expectations often result in conflict between the two, when indeed, the two should be allies, working together to resolve the matter.

When police officers and victims do work together, there is a much greater probability of resolving the case. Much of the conflict between police officers and victims stems from a lack of understanding on both parts. The police do not know what victims expect, need, or want from police officers, and victims do not know what police are expected to do.

This ambiguity and lack of clarification regarding the roles and responsibilities of police officers is enhanced as police departments move from a traditional style of policing toward a more "community" type of policing. It is difficult for police officers operating under a traditional style of policing to know what the expectations, in general, are for police officers. Without training or educating the officers, it will further separate police officers and victims.

The purpose of this book is to identify potential areas of conflict between police officers and victims. If we educate and train both current and future police officers on the issues of possible conflict between the two, we can begin to establish a more productive working relationship where the police and victims start to comprehend each others' position more accurately.

This reader is a collection of ten original chapters focusing on the topics of concern voiced by police officers. A focus group was conducted. About a dozen police officers with experience ranging from six to twenty years provided a list of topics they considered paramount for a victimology reader in order to increase their awareness and understanding of victims' concerns and issues.

Contributors to the reader are both practitioners in the field of criminal justice as well as academicians in the fields of criminology, criminal justice, law, policing, political science, social work, and public administration. The book is divided into three sections: Part I—Fundamental Issues, Part II—Victimization Issues, and Part III—Resource Issues.

Part I consists of three chapters. In the first chapter, Matt Robinson defines the term *victimology*. In the most elementary of definitions, victimology is the study of victims. However, as Robinson explains, the definition of victimology may focus on victims (i.e., people) harmed by illegal acts (e.g., personal or property crime) or it may be broadened to include other types of victims (i.e., organizations, groups, entities) harmed by illegal acts (e.g., environmental crime, corporate crime, white-collar crime, organized crime). Although other authors have argued for a broader definition of victimology, none have actually conceptualized the term. Robinson does this, and he also provides a typology of victimization by offender and victim type. Perhaps most important for this reader, Robinson takes his "new victimology" and applies it to policing—explaining how this new definition impacts policing.

In Chapter 2, Amie Scheidegger explains how the police should treat victims. As she points out in her chapter, "How the police respond to victims of crime impacts the outcome of a case, the recovery process of the victim(s), police-community relations, an individual's faith in and respect for the police, and his or her willingness to cooperate with police/law enforcement in the future." Therefore, it is imperative that police officers learn how to respond appropriately to crime victims. In her chapter, Scheidegger presents information regarding police-victim interactions to help police officers know what to expect from victims as well as what victims expect from the police. Further, she discusses how victims respond to crime, briefly discussing post-traumatic stress disorder. Scheidegger also examines distinct victim populations including elderly victims, victims with disabilities, and ethnically diverse victims and how the police should respond to these victims. The concept of secondary victimization is also discussed. Scheidegger concludes with the recommendation to develop and increase police officer training. ". . . the police must have a keen awareness of the various impacts crime has on victims. The officer must be well trained to understand how crime impacts the victim physically, psychologically, emotionally, and financially. An officer must understand the crisis process as it applies to victims and how to assist special victim populations. The more thoroughly trained an officer is, the more likely he or she will be able to respond adequately and appropriately to crime victims."

In Chapter 3, Peter Mercier addresses the subject of victim reporting—a topic discussed briefly in Chapter 2. As addressed by Mercier, "The police do not discover most crimes. Rather, the majority of crimes are reported to them." Therefore, it is essential that victims actually report crimes. Before making recommendations to increase victim reporting, Mercier provides a thorough review of the academic literature summarizing why victims do and do not report crimes. He then uses this information to develop strategies to increase victim reporting. An overlapping recommendation from Chapter 1 is the need for training of police officers, in particular, the need to train officers how to treat victims. Mercier makes the point that such efforts (and other efforts articulated in his chapter) will increase victim reporting.

Part II, Victimization Issues, contains three chapters. These chapters examine specific crimes and how the police should handle them. In Chapter 4, Tracy Woodard Meyers explains how the police should interview sexual assault victims. Dr. Meyers is a Florida Abuse Registry Counselor who brings her clinical experience and academic knowledge to her presentation on how to interview sexual assault victims. The police officers in the focus group were concerned about who should actually interview the victim. Should a female officer interview a female victim? Should a male officer interview a male victim? Should a female officer interview a child victim regardless of gender? Should there be more than one officer present as the interview is conducted? Should the interviewing officer be made a primary member of the investigation team? Dr. Meyers addresses these questions and more in her chapter, providing practical suggestions and instructions on how to interview victims of sexual assault.

In Chapter 5, Janet Hutchinson discusses the role of the police in investigating child welfare cases. These include dependent, neglected, and abandoned children, and children who have been physically and sexually assaulted. Dr. Hutchinson has consulted with state and local child welfare agencies in 48 of the 50 states over a 20-year period, and has a broad understanding of the issues involved in developing and maintaining successful multidisciplinary investigations of child abuse and neglect. Team members representing different disciplines bring very different perspectives of their responsibilities to the group effort—perspectives that often lead to misunderstandings that undermine multidisciplinary cooperation on child abuse and neglect cases. Dr. Hutchinson examines these goal conflicts, and the advantages and disadvantages of police involvement in child abuse and neglect cases. She also explores strategies that have been successfully used for developing and maintaining viable multidisciplinary teams that include police in investigations.

Chapter 6 focuses on domestic violence. Denise Kindschi Gosselin brings a unique perspective to this topic because she is a state trooper who has practical experience dealing with domestic disputes. She summarizes the academic literature on domestic violence providing the reader with an operational definition of domestic violence. She then discusses police intervention in domestic disputes. The main goal of the chapter is to provide officers with different ways to interview victims of domestic violence. Gosselin discusses purposeful interviewing and provides the reader with a set of misconceptions or false assumptions regarding what is expected at a domestic violence situation. This information is very important because erroneous expectations often lead to unacceptable or inappropriate behavior. By making the reader aware of these pitfalls, Gosselin is helping police officers respond to victims of domestic violence in a more effective manner. Her chapter continues with a full discussion of various interviewing techniques.

Part III, Resource Issues, contains four chapters. Each chapter examines a specific type of resource for victims. The police officers in the focus group were concerned they did not know enough about where to send victims for help. They expressed a desire to be more informed, in a general sense, about the services and resources available for victims.

In Chapter 7, Robert Jerin provides a historical and contemporary overview of victims' rights legislation. He details the grassroots effort that led to the adoption of the first victims' bill of rights. Although all states have adopted a statewide victims' bill of rights, currently there is no federal amendment to the U.S. Constitution. Jerin includes a list of rights that are found most often in a state's victims' bill of rights. Together these rights represent services. Police officers should be aware of what is promised to victims in such legislation.

In Chapter 8, Laura J. Moriarty and Robyn Diehl address state and federal resources available to victims. They begin with a short overview of why police officers should be concerned about crime victims' needs and what simple things can be done to address those needs. They discuss briefly what the police can expect from victims in terms of their reaction to crime and what victims expect from the police. In order to facilitate the recovery process, the authors advocate that police officers know about victim services on a national, regional, and local level. Moriarty and Diehl provide a list of resources on the federal level. They then specifically discuss one state recommending that the individual resources discussed be located in an officer's state and/or local community. Although compiling such a list will take a little effort on the part of the police department, it will ensure that the police have information to give to victims regarding available resources.

Chapter 9 focuses on one specific environment—the college campus. Max Bromley and Bonnie Fisher focus on campus policing and its evolution, as well as victim advocacy and victims' services provided on college campuses. They provide an overview of the extent and nature of on-campus crime. The authors provide detailed information regarding how colleges and universities have responded to the needs and mandates for victim services providing specific examples from Michigan State University and the University of South Florida. They conclude with a discussion of the challenges faced by campus police officials who provide services to campus crime victims, recommending strategies to reduce the barriers to quality service delivery.

Chapter 10 is the summary chapter. Dantzker and Moriarty discuss children as victims, providing an overview of the topic including a discussion of who protects children. The CASA program is detailed as well as special units in police departments. Finally, each chapter is briefly summarized.

This reader is intended as a resource for police officers, academicians and students. It may be used as a text for courses including Introduction to Criminal Justice/Criminology, any upper division policing course, or a victimology course or special topics course on victims of crime. More important, the reader should provide useful and timely information for both academic and police professionals interested in victims of crime.

Acknowledgments

I wish to thank all the contributors to this reader. I hope it has been a pleasant experience for each of you. This product is indeed a labor of love.

A special thanks to Mark Dantzker and Kim Davies for their support of this project.

Thanks also to my university, Virginia Commonwealth University, and my colleagues in the Dean's Office. Many times I worked on this book—pushing aside my other responsibilities. I know you all noticed but no one said anything. I really appreciate such a supportive environment.

Finally, I thank my family and friends. Just being able to discuss the contents of this book with others—not necessarily in the field—helped tremendously.

Contributors' Biographical Information

---❖---

Max L. Bromley is an associate professor in the Department of Criminology at the University of South Florida. He has previously served as the associate director of public safety at the University of South Florida and worked in the criminal justice field for almost 25 years. He is the senior coauthor of *College Crime Prevention and Personal Safety Awareness*, has coedited a volume entitled *Hospital and College Security Liability Awareness*, and is coauthor of the 5th edition of *Crime and Justice in America*. In addition, he has written numerous scholarly articles on campus crime and policing as well as technical documents on a variety of criminal justice topics.

Robyn Diehl received her bachelor's degree in psychology from Randolph-Macon College, her master's degree in criminal justice from Virginia Commonwealth University, and is currently enrolled in the doctoral program in developmental psychology at Virginia Commonwealth University, with an emphasis on the effects of community violence on the development of children. Her primary research interest is the effects of lethal and nonlethal violence on criminal behavior. Her most current research is in investigating the effects of violent crime on community response and participation with law enforcement. She has contributed to various technical reports, academic presentations, and scholarly articles.

Bonnie S. Fisher is an associate professor in the Division of Criminal Justice at the University of Cincinnati. She has a Ph.D. degree from Northwestern University (1988). Her research interests include student victimization, violence against college women, campus policing, institutions' responses to a report of sexual assault, and attitudes toward rehabilitation and corrections. Dr. Fisher's most recent work appears in *Criminology*,

Security Journal, Justice Quarterly, among others. She is the coeditor of the book, *Campus Crime: Legal, Social, and Policy Perspectives.*

Denise Kindschi Gosselin is a Massachusetts state trooper and an instructor in the Department of Criminal Justice at Western New England College. Her earned degrees include the master of science in criminal justice, Westfield State College (Massachusetts, 1990). Her research interests include domestic violence issues, distance learning, and juvenile justice. Trooper Gosselin is the author of *Heavy Hands: An Introduction to the Crimes of Domestic Violence* (Prentice-Hall, 2000).

Janet R. Hutchinson is an associate professor and the director of the Public Administration Program, Department of Political Science and Public Administration, Virginia Commonwealth University. Her earned degrees include the Ph.D. in public policy, University of Pittsburgh (1993). Dr. Hutchinson worked as a consultant and administrator in public child welfare agencies for 15 years before entering academia. Her most recent work has been in public policy decision making and knowledge use, and in applications of feminist theory to public policy decision making. She is currently working on a book that examines the development of family preservation policies in the United States between 1970 and 1990.

Robert A. Jerin is a professor and chair of the Law and Justice Department at Endicott College. His earned degrees include the Ph.D. in criminal justice from Sam Houston State University (1987). His research interests include restorative justice, crime prevention, victimology, and domestic violence. Dr. Jerin is the author of numerous book chapters and scholarly articles. He is the coauthor of *Victims of Crime* (Nelson Hall) and the coeditor of *Current Issues in Victimology Research* (Carolina Academic Press).

Peter J. Mercier is a special agent with the Naval Criminal Investigative Service, specializing in computer-related crimes. He has 17 years of law enforcement experience. His earned degrees include the master of art in sociology, Old Dominion University. Professor Mercier is an adjunct instructor at Old Dominion University and St. Leo College. His research interests include domestic violence issues and computer deviance. Professor Mercier is the coeditor of *Battlecries from the Homefront: Violence in the Military Family* (Charles C. Thomas, 2000).

Tracy Woodard Meyers is an associate professor in the Department of Sociology, Anthropology, and Criminal Justice at Valdosta State University. She received the Ph.D. in family relations with an emphasis in traumatology from Florida State University (1996). She has conducted research in the area of secondary traumatic stress. Her research interests include domestic violence, crisis intervention, child welfare, and trauma and the family. Dr. Meyers is a Florida Abuse Registry Counselor. Additionally, before joining the faculty at VSU, Dr. Meyers spent 10 years working with trauma victims in a variety of social service agencies. She is the author of numerous research articles and book chapters.

Laura J. Moriarty is a professor of criminal justice and assistant dean, College of Humanities and Sciences, at Virginia Commonwealth University. Her earned degrees include the Ph.D., Sam Houston State University (1988). Her research areas include victims of crime, victimology, fear of crime, and violent crime. She is the author, coauthor, or coeditor of four books: *Victims of Crime* (with Robert Jerin, Nelson-Hall, 1998), *American Prisons: An Annotated Bibliography* (with Elizabeth McConnell, Greenwood Press, 1998), *Current Issues in Victimology Research* (with Robert Jerin, Carolina Academic Press, 1998), and *Criminal Justice Technology in the 21st Century* (with David Carter, Charles C.

Thomas, 1998). Additionally, Dr. Moriarty has published numerous scholarly articles, book chapters, and nonrefereed articles.

Matthew B. Robinson is an assistant professor of criminal justice in the Department of Political Science and Criminal Justice at Appalachian State University. His earned degrees include the Ph.D. in criminology and criminal justice, Florida State University (1997). His research areas include criminological theory, criminal victimization, and crime prevention. Dr. Robinson is the author of many scholarly articles appearing in the *American Journal of Criminal Justice, Journal of Contemporary Criminal Justice, Howard Journal of Criminal Justice, Corrections Compendium, Western Criminology Review,* and *Journal of Crime and Justice.*

Amie R. Scheidegger is an assistant professor of criminal justice at Charleston Southern University. She received her B.S. in criminal justice from Illinois State University in 1990. She earned her M.S. in 1993 and Ph.D. in 1998 in criminology/criminal justice from Florida State University. Her major research interests are female crime and crime prevention. She is currently researching in the area of domestic violence.

1

The Case for a "New Victimology"

Implications for Policing

Matthew B. Robinson

INTRODUCTION

Imagine that you buy a car from a used car dealer in your town. The car is in nice physical shape, almost like new except for the deep scratches in one of the front fenders.
Since the salesperson of the car dealership assures you that the scratches can easily be
repaired by their own body shop, you decide to buy the car. His words are, "We can make
it look like brand new." After filling out and signing all of the required forms, including
the sales contract, the owner of the dealership instructs you to come back in one week
after the repairs are completed to take possession of your car.

Upon retrieving your car, you immediately notice that the scratches are still visible. Although the damage is not as severe as it originally was, the car hardly looks
"brand new" as you were verbally promised by the salesperson. You get angry and demand to see the owner of the dealership. You insist that the dealership attempt to repair
the car again since the salesperson told you the scratches would not be visible once the
repairs were completed. The owner of the dealership tells you that there is nothing she
can do to appease you because you signed the paperwork and nothing in the contract
noted the damage to the vehicle. She points out that the contract even states that any verbal promises are not considered as valid parts of the contract.

So, you are left with little choice but to drive your still-damaged car home. You
feel sheepish and naive, and you are sick to your stomach. You feel like you have been
victimized by the dealership. But have you? Are you a victim? What would victimologists say?

1

Victimology can be generally understood as the study of victims of crime. Although numerous sources define the concept differently (e.g., see Doerner & Lab, 1998; Jerin & Moriarty, 1998; Karmen, 1996; Kennedy & Sacco, 1998; Moriarty & Jerin, 1998; Sgarzi & McDevitt, 2000; Wallace, 1998), victimology is a subfield within the discipline of criminology (Champion, 1997, p. 128), having grown out of other disciplines such as law, psychology, and sociology. As defined by Rush (2000, p. 336), victimology is the "study of the psychological and dynamic interrelationships between victims and offenders, with a view toward crime prevention."

Wallace (1998, p. 3) defines victimology as "the study of the victim, the offender, and society" which "encompasses both the research and scientific aspects of the discipline as well as the practical aspects of providing services to victims of crime." Victimologists are interested both in why and how people become victims, as well as how victims are and should be involved in the criminal justice process (Champion, 1997). Karmen (1996, p. 2) defines victimology as "the scientific study of crime victims [which] focuses on the physical, emotional, and financial harm people suffer at the hands of criminals." As such, victimologists study aspects of the victim-offender relationship, including but not limited to the role victims play in criminal events, as well as "the public's reaction to the victim's plight, the criminal justice system's handling of victims, and victims' attempts to recover from their negative experiences" (Karmen, 1996, p. 2).

MAJOR ISSUE: AN ALTERNATIVE CONCEPTION OF VICTIMOLOGY AND ITS IMPLICATIONS FOR POLICING/LAW ENFORCEMENT

The concept of victimization, and thus the field of victimology, has been applied almost exclusively to crime victims who were victimized by individuals (i.e., crimes committed by individuals against other individuals). This means that your victimization at the hands of the car dealership described earlier would not likely be of concern to most victimologists. An alternative conception of victimization and definition of victimology would allow for the study of other forms of harmful behavior, some of which are criminal and some of which are not, some of which are committed by individuals against other individuals and some of which are not. The scientific study of the other harmful behaviors that have traditionally been ignored by victimologists—even though they produce victims—would have significant implications for criminal justice agencies, including police/law enforcement agencies.

This chapter concerns itself with whether the field of victimology should be expanded to other areas of victimization. I argue that it should, by proposing a new definition of victimology and then by addressing the rationale for this argument. Finally, I discuss the effects that expanding the scope of victimology will have on policing/law enforcement.

WHAT IS VICTIMIZATION? THE TRADITIONAL VIEW IN CRIMINOLOGY AND VICTIMOLOGY (AND ALTERNATIVE VIEWS)

Victimization is a measure of the occurrence of a crime, whereby an individual person becomes a victim (suffers financial or physical harm) as a result of a criminal act. Rush (2000, p. 335, italics added) writes that victimization is "the harming of any *single vic-*

tim in a *criminal* incident." Champion (1997, p. 128, italics added) defines victimization as a "specific *criminal* act affecting a *specific victim.*" Notice the common elements of the definitions:

- A single victim;
- Suffering harm;
- From a criminal act.

According to these definitions, victimization occurs when one person suffers some harm from a behavior that violates the criminal law. Yet, people obviously become victims of behaviors that do not meet the conditions of the preceding definitions. Table 1-1 sets forth a topology of victimization that illustrates four basic types of victimization. These include:

1. Harmful acts committed by an individual person against another individual person (Cell 1 in the table);
2. Harmful acts committed by an individual person against an entity (e.g., a group of people or a corporation) (Cell 2);
3. Harmful acts committed by an entity (e.g., a group of people or a corporation) against an individual person (Cell 3); and
4. Harmful acts committed by an entity (e.g., a group of people or a corporation) against an entity (e.g., a group of people or a corporation) (Cell 4).

Table 1-1 also provides examples of specific types of behaviors that fall into each cell. Examples of harmful acts committed by an individual person against another individual person include theft (e.g., taking property from another without consent and without use of force or threat of force) and battery (e.g., physically striking or beating another person with force). Examples of harmful acts committed by an individual person against a group of people or a corporation include forgery (e.g., signing another person's name to a check that does not belong to you in order to cash it) and terrorist acts (e.g., planting and detonating explosives at a building occupied by other people). Examples of harmful acts committed by a group of people or a corporation against an individual person include fraud (e.g., making false or deceptive claims about a product in order to trick the buyer into purchasing the product) and manufacturing and selling defective products (e.g., making a car that explodes in a low-impact crash due to a faulty part). Finally, examples of harmful acts committed by a group of people or a corporation against a group of people or a corporation include price fixing (e.g., when two corporations keep prices artificially high by agreeing to set a specific price on their products and simultaneously leave out smaller competitors who thus cannot lower prices to compete for buyers' business) and forcing or allowing employees to work in hazardous conditions (e.g., not complying with federally required safety regulations or ignoring warnings of potential problems).

It should be noted that although these four types of victimizations are mutually exclusive, the examples provided in Table 1-1 are not. That is, some of the examples provided could also fit well into another category of victimization. For example, acts of fraud can also be committed by an individual against another individual (such as when someone sells a fake fur coat to someone after claiming it is "a genuine mink coat") or by a group of people or a corporation against an individual person (such as when a corporation sells a product that will "make you younger!"). Likewise, theft could also be

TABLE 1-1   A Topology of Victimization by Offender and Victim Type

		Offender	
		Individual	Group/Corporation
	Individual	Cell 1	Cell 3
		Theft	Fraud
		Battery	Defective product
Victim			
	Group/Corporation	Cell 2	Cell 4
		Forgery	Price fixing
		Terrorist act	Dangerous working condition

committed by individuals against individuals or groups and corporations. The examples simply serve to assist with understanding the four different types of victimizations defined previously.

It should also be pointed out that because some forms of victimization may stem from acts that are criminal (i.e., in violation of the criminal law) and some are simply deviant (i.e., in violation of some standard or prevailing norm or expectation for acceptable behavior) but not illegal, there are actually eight types of victimization. This is because each of the four types of victimization in Table 1-1 can either be illegal or legal. For purposes of this analysis, we will only address the four types of victimization discussed earlier in this chapter.

The most significant point for this chapter is that victimization encompasses so many more types of behaviors than can be adequately understood according to the most adhered to definition in victimology. As noted by Karmen (1996, p. 2, italics added), the term *victim* most accurately "refers to all those people who experience injury, loss, or hardship due to *any cause*." In other words, a broader definition of victimology would be:

> the study of people who experience any harmful behavior, such as "accident victims, cancer victims, flood victims, and victims of discrimination and similar injustices" and people who have been "physically injured, economically hurt, robbed of self-respect, emotionally traumatized, socially stigmatized, politically oppressed, collectively exploited, personally alienated, manipulated, co-opted, neglected, ignored, blamed, defamed, demeaned, or vilified" (Karmen, 1996, p. 2).

Karmen's very general definition could encompass many direct and indirect types of victimizations. Realistically, it may be too general to assist in sustaining this scientific field of study. Given that such a broad conception of victimology would allow and even require victimologists to study virtually any behavior, there would likely be little systematic and continued study of any particular type of harmful behavior. So, perhaps a more restrictive definition is in order.

A "NEW VICTIMOLOGY" FOR CRIMINOLOGY AND CRIMINAL JUSTICE

To the degree that victimology is a subset of criminology, it will likely be focused on acts that are already considered illegal—that is, crimes. This definition is too restrictive to encompass the wide range of purposive behaviors that create victims, particularly given the fact that "crime" is a label applied by persons with power to simultaneously have their will enacted into law and to create a criminal law not focused on their own harmful and deviant acts (e.g., see Robinson, 1998). A compromise between the traditional definition of victimology and the broader one proposed earlier would allow victimologists to study a specific set of harmful behaviors that more accurately capture the types of culpable behaviors which cause harm to their victims. The key word pertaining to criminology is *culpability,* which indicates some degree of blameworthiness or responsibility for a resulting action. Culpable actions include those that are committed:

- Intentionally (i.e., committed with a guilty mind on purpose);
- Negligently (i.e., committed as a result of a failure to meet normal or recognized expectations);
- Recklessly (i.e., committed without due caution for human life or property); and
- Knowingly (i.e., committed with knowledge).

I suggest that the concept of victimization be broadened to include *any act that produces financial or physical harm and that is committed intentionally, negligently, recklessly, or knowingly.* A similar point is made by Williams (1996, p. 19, italics in original) when discussing "environmental victims" in a special issue of *Social Justice:* "the concept of 'victims' embodies the idea of suffering caused by a *deliberate or reckless human act* [including an act of omission]." Williams argues that for a person to be held accountable for an act against the physical environment, "the outcome of an act must have been 'reasonably foreseeable' for it to constitute an offense."

Following this logic, a "new victimology" is in order, one that consists of *the scientific study of individual persons, groups of people, or corporations who are victimized by acts committed intentionally, negligently, recklessly, or knowingly.* This is very similar to the notion of a "radical victimology" proposed by scholars such as Friedrichs (1983), Phipps (1986), and McShane and Williams (1992).

Radical victimology does not have a clear definition, per se, but those who have called for it have been critical of traditional victimology for at least the following reasons:

- Current images of victims reinforce a focus on the state rather than the actual person(s) who suffered physical or financial harm. This "is a logical extension of a legal system which defines crimes as offenses against the state" (Zehr & Umbreit, 1982, p. 64);
- Current images of the state as victim reinforce a "conservative crime control agenda" or ideology "and have increased the power of the state in criminal proceedings" (McShane & Williams, 1992, p. 258); and
- Victimology is focused on street crimes of the poor rather than white-collar crimes and deviance which allows such offenders to escape relatively unscathed and unpunished despite their culpability (e.g., see Simon & Hagan, 1999).

Such criticisms have led scholars to call for a widening of the scope of victimology beyond the study of street crimes committed by individuals against other individuals—that is, "to extend victimology into neglected areas" (McShane & Williams, 1992, p. 267). In essence, I have made the same argument in this chapter, but in so doing have actually proposed a definition of this "new victimology," something not yet done by any other author(s).

Before turning to the implications of this "new victimology" for police/law enforcement, one final note is in order. One might reason that, through this offering of a "new victimology," I am advocating an expansion of the definition of crime beyond purely legal conceptions. Criminalization of behaviors is a complex process involving numerous actors. The law serves some very clear and limited interests at the expense of other groups (mainly poverty-stricken minority males and small-time drug offenders) (Robinson, in press). Yet, an expanded conception of "crime" is not necessary for a "new victimology." Instead, as others have argued (e.g., see Krisberg, 1975; Quinney, 1972; Schwendinger & Schwendinger, 1970; Taylor, Walton, & Young, 1972), I am simply calling for an increased recognition of, as well as study of, other types of victims besides those typically examined by victimologists. Perhaps this type of research will demonstrate once and for all that far more people are killed and physically or financially injured by acts committed intentionally, negligently, recklessly, or knowingly which are *legal* than which are *criminal*. This ultimately may lead to criminalization of certain forms of behaviors that are currently legal. Once an act becomes criminal, its victims require services from the criminal justice system, particularly police/law enforcement, the agencies with the most direct contact with citizens.

More important for the survival of victimology is that in order to accurately do what victimologists are supposed to do, we must first be able to identify, define, and describe the problem of victimization and measure its true dimensions—only then can we investigate how the criminal justice system handles all victims and learn how society might respond to their victimization (Karmen, 1996). As laid out in the United Nations' *Declaration of Basic Principles of Justice for Victims of Crime and Abuse of Power* (1985), far fewer people have become victims of crime than of abuse of power by oppressive governments and what Petrovec (1997) calls "redress"—which includes the steps of newly empowered oppressed groups to get even with those who oppressed them. The point seems already clear to criminologists that the "image of victims of ordinary offenses" such as theft and burglary (Petrovec, 1997, p. 164) is misleading (e.g., see Friedrichs, 1996; Reiman, 1998; Rosoff, Pontell, & Tillman, 1998; Simon & Hagan, 1999). Victimology has yet to have learned this lesson.

IMPLICATIONS FOR POLICING/LAW ENFORCEMENT OF A "NEW VICTIMOLOGY"

Now I turn to the issue of how broadening the scope of victimology will affect policing/law enforcement in America. I first discuss how policing/law enforcement is organized in the United States, in order to demonstrate at what level of government most police/law enforcement activity occurs. Second, I examine basic roles and responsibilities of police/law enforcement officers in the United States, in order to demonstrate what the "typical police/law enforcement officer" (Wrobleski & Hess, 2000) does during his or her shift. By examining these basic police/law enforcement issues, I illustrate two important points that have implications for victimization and thus victimology:

1. Policing/law enforcement in America is organized in a manner which mandates that officers focus on street crime victimizations almost exclusively; and
2. Police/law enforcement officers in America spend most of their time doing work not related to crime or victimization.

Given these facts, a reorganization of police/law enforcement resources (including the numbers of police/law enforcement officers allocated to each level of government, the types of behaviors they focus on, and the ways in which they use their time) may be in order, so that victims of all culpable harmful acts, whether legal or illegal, will be served by police/law enforcement. This would make police/law enforcement activities consistent with a "new victimology."

ORGANIZATION OF POLICING/LAW ENFORCEMENT IN AMERICA

The allocation of police/law enforcement officers to various government levels (i.e., city, county, state, federal governments) in America will have major effects on what police/law enforcement do. Since policing/law enforcement at different levels of government is focused on different forms of behaviors, if we examine at what level of government the police/law enforcement work in the United States, we can get a *general* sense of their activities. Table 1-2 demonstrates quite conclusively that most police/law enforcement activity occurs at the local level (i.e., city or county government). As shown in Table 1-2, 56% of sworn officers are city police/law enforcement officers, followed by 21% county sheriff officers. This means that 77% of sworn officers are local police/law enforcement officers. Another 13% are state level officers and 10% are federal officers. This distribution results in a rate differential of 2.5 local and state police/law enforcement officers per 1,000 citizens versus 0.28 federal officers per 1,000 persons (United States Department of Justice [USDOJ], 1998).

Table 1-3 shows that most U.S. police/law enforcement departments are local level agencies (13,578), followed by county sheriffs' departments (3,088), and specialized police/law enforcement departments (1,316). There are currently only 27 federal police/law enforcement agencies with an employment of at least 100 agents. Overall, local and state agencies employed 921,978 people on a full-time basis in 1996, including 663,535 full-time sworn officers with general arrest powers (72%) and 258,443 nonsworn officers or civilians (28%) (USDOJ, 1998).

TABLE 1-2   Number of Sworn U.S. Police/Law Enforcement Officers by Level of Government and Percent of Total (1996)

	Level of Government				
	Local				
	City	County	State	Federal	Total
Number of Officers	410,956	152,922	99,657	74,493	738,028
Percent of Total	56%	21%	13%	10%	100%

TABLE 1-3   Number of U.S. Police/Law Enforcement Agencies
              by Level of Government (1996)

	Level of Government	
	Local/State	Federal
Local Departments	13,578	—
Sheriff Departments	3,088	—
Special	1,316	—
Texas Constables	738	—
Primary State Agencies	49	—
Total	18,769	27 (including agencies with more than 100 agents)

As noted by the USDOJ (1998, p. 3): "general purpose local police/law enforcement departments were the largest employer with 521,985 full-time employees as of June 1996."

Although local police/law enforcement in the United States direct most of their crime-fighting efforts at relatively harmless acts that are not included in the Part I Index Offenses of the Uniform Crime Report (UCR), their general focus is still on street crime. This means that even though they are more likely to pursue petty criminals, those offenders who are considered to pose the greatest threats to the community are those who have committed one of the eight "most serious" Part I Index Offenses of the UCR. These acts include theft, burglary, motor vehicle theft, arson, homicide, aggravated assault, forcible rape, and robbery. These acts are generally perceived by society and government agencies alike to be the behaviors that cause the most physical and financial harm and that occur with the greatest frequency (Robinson, 1999). As pointed out by Wrobleski and Hess (2000, p. 132): "Usually [police/law enforcement] departments concentrate police/law enforcement activities on serious crimes—those that pose the greatest threat to public safety and/or cause the greatest economic losses." The main conclusion from the figures in Tables 1-2 and 1-3 is that the general focus of police/law enforcement in the United States is on a very small amount of criminal acts, and an even smaller amount of culpable harmful acts. This means that the majority of types of victimizations, even those that stem from harmful, culpable behaviors, are virtually ignored by police/law enforcement in America.

Figure 1-1 illustrates this point, and shows that police/law enforcement in the United States is most focused on criminal victimizations of individuals caused by acts of individual persons. The shaded region in Figure 1-1 indicates the main focus of police/law enforcement when it comes to victimizations. Policing/law enforcement is much less concerned with victimizations against individuals that are committed by entities such as groups or organizations. Such acts are usually pursued by state or federal level police/law enforcement agencies. Victimizations against individuals or entities such as groups and corporations committed by other entities are *generally not handled* by the criminal justice system; hence they do not fall within the primary domain of police/law enforcement. Instead,

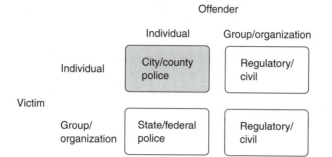

FIGURE 1-1 Focus of law enforcement by type of victimization.

harms and victimizations resulting from such acts are handled through civil law (as in the case of major lawsuits against large tobacco corporations) or regulatory agencies (as in the case of outbreaks of bacterial contamination in food products). Other harmful culpable behaviors, especially those that do not violate any law, are completely ignored.

BASIC ROLES AND RESPONSIBILITIES OF POLICE/LAW ENFORCEMENT OFFICERS

According to Wrobleski and Hess (2000, p. 128), police/law enforcement officers in America have five basic goals, including:

1. Enforcing laws—including investigating reported crimes, collecting and protecting evidence from crime scenes, apprehending suspects, and assisting the prosecution in assuring convictions;
2. Preserving the peace—including intervening in noncriminal conduct at public events and places which could escalate into criminal activity if left unchecked;
3. Preventing crime—including numerous forms of activities to stop crime before it occurs, such as education campaigns, preventive patrols, community policing, and so on;
4. Providing services—including performing functions normally served by other social service agencies, such as counseling, referring citizens for social services, assisting people with needs, keeping traffic moving, and so on; and
5. Upholding civil rights—including respecting all persons' rights regardless of race, ethnicity, class, gender, and so forth, and reading Miranda warnings to suspects when arrest is imminent.

Of these five roles, the "typical police/law enforcement officer" spends most of his or her time each day *not fighting crime*—"approximately 90% of a police/law enforcement officer's time is spent in the social service function" (Wrobleski & Hess, 2000, p. 128). As noted by Manning (1997, p. 93): "Of the police/law enforcement functions or activities most central to accumulated police/law enforcement obligations, none is more salient than supplying the range of public services required in complex, pluralistic, urban societies." So, although "crime fighter" is the stereotypical view of the police/law enforcement officer, the typical police/law enforcement officer in the United States (i.e., the city patrol officer or county sheriff) spends the smallest amount of his or her day assisting victims of crime.

An examination of activities of police/law enforcement officers at various levels of government illustrates this point equally well. Table 1-4 shows the main function served by U.S. police/law enforcement officers by level of government in 1996. As shown in Table 1-4, most officers have primary responsibility for answering calls for service (64%), followed by investigative duties (15%), administrative/technical/training duties (10%), jail-related duties (8%), and court-related duties (3%). Of the sworn officers serving local police/law enforcement agencies, 70% were involved in patrol and responding to calls for service, versus 16% who had investigative duties, 12% with administrative/technical/training duties, and 2% with jail and court duties (USDOJ, 1998). County sheriff officers had similar duties, including 42% involved in patrol and responding to calls for service, 30% with jail duties, 12% with investigative duties, 11% with court duties, and 5% with administrative/ technical/training duties (USDOJ, 1998). The main difference between local police/law enforcement and county sheriffs' officers is that county sheriffs' officers are more heavily involved in jail and court duties.

State officers were most assigned to responding to calls for service (69%), followed by investigative duties (15%), administrative/technical/training duties (14%), and court-related duties (2%). Table 1-4 also demonstrates that 56% of federal officers have primary responsibility for investigations and enforcement (43% for criminal matters), followed by 21% for corrections-related duties, 16% for police/law enforcement services, 4% for court operations, and 3% for security and protection. The majority of federal officers (58%) work for the Immigration and Naturalization Service (12,403), the Bureau of Prisons (11,329), the Federal Bureau of Investigation (10,389), and the U.S. Customs Service (9,749) (USDOJ, 1997). While most INS workers are border patrol agents, FBI officers have broad investigative responsibilities for more than 250 federal crimes such as bank fraud, embezzlement, kidnapping, and so forth.

Given that only one in two crimes is even known to the police/law enforcement (and this only counts street crimes), and only one in five of these leads to an arrest (Cole & Smith, 1998), it should not be surprising that police/law enforcement have some free time on their hands. Therefore, I would advocate a redistribution of policing/law enforcement resources to better assist victims of all culpable, harmful acts. Police/law enforcement should devote free time to other forms of victimization than street crimes committed by individuals against individuals. We might also decide as a society to let police/law enforcement fight victimization exclusively, and leave service functions to other agencies. Why not redesign our social service agencies to operate on a 24-hour basis, so that police/law enforcement may exclusively fight victimization and/or crime through their law enforcement function? Given the amount of time police/law enforcement at all levels of government spend on service functions, it is apparent that a great deal of time is not being allocated to victim assistance, although it is difficult to ascertain how much of police/law enforcement service time is spent on precisely this function. Additionally, many criminal victimizations are never discovered and virtually all noncriminal victimizations are ignored. From these facts, we can confidently conclude that police/law enforcement are not very effectively serving victims of culpable harms. A "new victimology" would promote change in this area.

TABLE 1-4 Main Function Served by U.S. Police/Law Enforcement Officers by Level of Government (1996)

	Level of Government	
	Local/State	Federal
Calls for Service	64%	16%
Investigative Duties	15%	56%*
Administrative/ Technical/Training	10%	0%
Jail-Related Duties	8%	21% (corrections)
Court-Related Duties	3%	4%
Security/Protection	0%	3%
Total	100%	100%

*Of federal officers involved in investigation, 43% were involved in criminal investigation and enforcement and 13% were involved in noncriminal investigation and enforcement.

IMPLICATIONS OF THE "NEW VICTIMOLOGY" FOR POLICING

So then, what would be the specific implications of this "new victimology" for American policing/law enforcement? As argued earlier, the first would be a redistribution of police/law enforcement time and resources to other forms of victimization. In light of the available evidence regarding the relative harms associated with legal versus illegal behaviors, it is clear that legal forms of behavior—what Robinson (1999) refers to as "excluded harms" because they are generally excluded from criminal justice system attention—cause much more physical and financial harms than do street crimes. Given this fact, it is illogical to tolerate the differential focus by police/law enforcement on street crimes, which are minor *relative to other forms of culpable harmful behavior.*

Criminologists known as "left realists" (e.g., see Schwartz & DeKeseredy, 1991) correctly point out that street crimes are very harmful and thus should not be ignored. This point is well-taken, for many forms of street crime are particularly heinous and troubling. For example, domestic or intimate violence occurs between family members and people who know each other. As of June 1997, 83% of local police/law enforcement departments and sheriffs' agencies in America had written policies on how to properly handle domestic disputes (USDOJ, 1999). Policies on handling other types of victims of illegal and legal culpable harmful acts should also be developed and implemented. For example, when FBI officers investigate crimes such as bank fraud, embezzlement, and kidnapping, do they have policies on how to properly handle victims of such crimes? What about victims of false advertising and price gouging, who generally have nowhere to turn for assistance from police/law enforcement (Friedrichs, 1996)?

Many of these victims, as well as those who are physically maimed or injured at work also receive little attention from police/law enforcement agencies. As noted by Friedrichs

(1996, p. 271), "the proportion of apparent white collar crimes that are officially investi-gated and lead to enforcement actions is lower than is the case for conventional crime. In the simplest and most colloquial terms, what occurs in the street is more visible and more easily investigated than what occurs 'in the suite,' behind closed doors." But, just because it is hidden, does this mean its victims are not worthy of both compensation for their suf-fering and justice for the harms committed against them?

Such victims are typically left with assistance from state and federal police/law en-forcement agencies because of the "complex, often interjurisdictional character" of the harmful acts which produce the victimizations, and/or regulatory agencies (Friedrichs, 1996, p. 272), as depicted in Figure 1-1. Police/law enforcement agencies involved in-clude "over two dozen separate federal agencies" including "the Federal Bureau of In-vestigation (FBI), the Inspector Generals, the U.S. Postal Inspection Service, the U.S. Se-cret Service, the U.S. Customs Service, and the Internal Revenue Service Criminal Investigative Division" (Friedrichs, 1996, p. 273). We've already seen how few officers work for these agencies relative to city and county police/law enforcement forces in the United States. Regulatory agencies include the Occupational Safety and Health Admin-istration (OSHA) for negligent workplace injuries and deaths, the Food and Drug Ad-ministration (FDA) for adulterated food products, the Consumer Product Safety Com-mission (CPSC) for unsafe products, and the Environmental Protection Agency (EPA) for acts against the environment. According to Friedrichs (1996, p. 285), "regulatory agen-cies are greatly understaffed and underfunded . . . Public pressure for agency action is small relative to that for conventional crime, and business interests have traditionally lob-bied for various limitations on agency powers and budgets." For example, OSHA only has hundreds of inspectors who must investigate millions of businesses. Although victims of culpable harms committed by entities such as groups of people or corporations can also turn to the civil law for redress, it is clear that their victimizations are not treated as seri-ously as street crimes.

Ironically, while such victims are generally ignored by local police/law enforce-ment agencies, our police/law enforcement officers spend a disproportionate amount of time and resources on some "victimless" street crimes, called this not because they are truly harmless but because they are engaged in by consenting adults so that no one in-volved feels victimized. The best example I can think of, primarily because of how much criminal justice money is spent on it, is the "war on drugs." As explained by Robinson (in press), factors such as the police/law enforcement focus on drug crimes and manda-tory sentencing laws that call for longer minimum sentences for drug offenders, explain America's unprecedented overreliance on imprisonment. Many of the people we are incarcerating are first-time, low-level drug dealers or people who simply were found in possession of marijuana. Meanwhile, victims of real forms of harms are virtually ignored.

We can easily understand why most policing/law enforcement in the United States is focused exclusively on street crimes instead of other harmful acts committed intentionally, recklessly, negligently, or knowingly. Friedrichs (1996, p. 271) claims this owes itself to a lack of jurisdiction, expertise, and resources. To facilitate police/law enforcement involve-ment in victimizations caused by other culpable harmful acts, changes must be made with regard to police/law enforcement jurisdiction, expertise, and resources.

HOW POLICE/LAW ENFORCEMENT CURRENTLY SERVE "VICTIMS": IMPLICATIONS FOR THE "NEW VICTIMOLOGY"

If local police/law enforcement departments are not highly involved with victims of harmful acts other than street crimes, how do police/law enforcement currently serve crime victims? And how could this be changed to bring local police/law enforcement activities in line with a "new victimology"? Generally, as police/law enforcement "are the first representatives of the criminal justice system that victims encounter in the immediate aftermath of crimes," they serve victims in the following ways:

- Responding quickly to calls for help;
- Launching thorough investigations into alleged crimes;
- Preserving and collecting evidence;
- Solving crimes by capturing suspected offenders; and
- Assisting with criminal prosecutions of criminal suspects (Karmen, 1996, p. 166).

A "new victimology" based on an expanded conception of victimization would require police/law enforcement to service new forms of diverse types of victims (e.g., victims of culpable acts committed by or against entities such as groups of people or corporations) in a similar manner. This would mean that police/law enforcement would be responding quickly to different types of calls for help, launching thorough investigations into different forms of alleged crimes, preserving and collecting evidence of those offenses, solving crimes by capturing different types of alleged offenders, and assisting in their prosecutions.

In order for this to happen, at least three changes to American policing must accompany the "new victimology." These would include:

1. Reallocating police/law enforcement resources from city and county agencies to state and federal agencies who currently focus on the other forms of victimization;
2. Reeducating police/law enforcement officers about forms of victimization other than street crimes; and
3. Developing advertising campaigns aimed at educating the general public about forms of culpable harms.

First, as discussed previously, state and federal officers are best equipped to provide assistance to victims of other forms of victimizations, given their special training. In order to assure that victims of such acts are sufficiently attended to, police/law enforcement at the state and federal levels must be granted more resources. This means, relative to local police/law enforcement, state and federal police/law enforcement agencies should be granted more employees and financial resources.

Second, police/law enforcement officers at the local level must be educated about forms of culpable harms other than street crimes, particularly to the degree that they will likely be called to assist with their victims. City and county police/law enforcement assistance would not be possible unless allocation of local police/law enforcement personnel by location and time would be systematically changed from urban areas where street crime is perceived to be occurring to areas where other forms of culpable harms occur. That is, for local police/law enforcement to provide their traditional services to a more diverse set of victims, where and when they patrol would also change, because as "police/law enforcement engage in routine patrol work, they may encounter individuals or situations they define as criminal" (Sacco & Kennedy, 1996, p. 64). Since police/law enforcement "tend to respond most emphatically to . . . events that they perceive as conforming to legal definitions of serious crimes" (Sacco & Kennedy, 1996, p. 65), they would be unlikely to sympathize with victims of culpable behaviors other than street crimes, unless such acts were recognized as serious crimes by society and treated as such by the criminal justice system.

Third, since policing/law enforcement will likely remain largely reactive in nature and will continue to rely on victims for assuring police/law enforcement success in solving cases, victims must first report their victimizations to police/law enforcement. As noted by Karmen (1996, p. 178): "The single most important factor in solving crimes . . . is the ability of victims to furnish detectives with clues, leads, and descriptions, or even the names, of suspects." For this to occur, we must educate citizens about their true risks of victimization. Victims of other forms of culpable harmful acts, such as victims of white-collar crimes, "are often much more confused than are victims of conventional crime about where to turn for help, and a much larger group of such victims are not even conscious about having been victimized" (Friedrichs, 1996, p. 277). If we truly want to stop victimization, we must make sure people know they are being victimized, and then increase and clarify their options for police/law enforcement assistance.

CONCLUSION

In this chapter I have demonstrated that most victims of culpable harmful acts are not being served by American police/law enforcement officers. Typically, police/law enforcement agencies limit their services to criminal acts committed by individuals against other individuals. This owes itself to the way in which policing/law enforcement is organized in America, where most resources and personnel are allocated to the local (i.e., city and county) level. Since local police/law enforcement primarily focuses on street crimes, other forms of harmful acts committed intentionally, negligently, recklessly, and knowingly are left up to state and federal police/law enforcement or regulatory agencies, which are understaffed and overburdened. Victims of such acts generally must turn to the civil law for redress.

In order to right this wrong—that is, to assure that more victims get justice for the harms committed against them by culpable actors—I put forth a "new victimology," one that consists of *the scientific study of individual persons, groups of people, or corporations who are victimized by acts committed intentionally, negligently, recklessly, or knowingly.* This chapter serves as a call to action for those who consider themselves victimologists and for those who simply study victimization, to expand the scope of their study beyond legalistic conceptions of crimes (especially street crime).

In order for American police/law enforcement operations to be consistent with this "new victimology," I suggested the following changes to the way policing/law enforcement in America is done: (1) lessening the focus of police/law enforcement on street crimes, particularly victimless crimes; (2) emphasizing other forms of culpable harms which currently are being ignored; (3) reallocating police/law enforcement resources from city and county agencies to state and federal agencies who currently focus on the other forms of victimization; (4) reeducating police/law enforcement officers about forms of victimization other than street crimes; and (5) developing advertising campaigns aimed at educating the general public about forms of culpable harms. This undoubtedly is a radical call for change, but it is one that must be answered if the criminal justice system is to be truly just. The current course of criminal justice in America is too heavily focused on victims of acts committed by a very small segment of the population. Other victims of harmful acts are being virtually ignored for no reason other than history or tradition.

REFERENCES

CHAMPION, D. (1997). *The Roxbury dictionary of criminal justice*. Los Angeles: Roxbury.

COLE, G., & SMITH, C. (1998). *The American system of criminal justice* (8th ed.). Belmont, CA: Wadsworth.

DOERNER, W., & LAB, S. (1998). *Victimology* (2nd ed). Cincinnati, OH: Anderson.

FRIEDRICHS, D. (1983). Victimology: A consideration of the radical critique. *Crime & Delinquency, 29*(2), 283–294.

FRIEDRICHS, D. (1996). *Trusted criminals: White collar crime in contemporary society*. Belmont, CA: Wadsworth.

JERIN, R.A., & MORIARTY, L.J. (1998). *Victims of crime*. Chicago: Nelson Hall.

KARMEN, A. (1996). *Crime victims*. Belmont, CA: Wadsworth.

KRISBERG, B. (1975). *Crime and privilege*. Englewood Cliffs, NJ: Prentice-Hall.

KENNEDY, L.W., & SACCO, V.F. (1998). *Crime victims in context*. Los Angeles: Roxbury.

MANNING, P. (1997). *Police work: The social organization of policing*. Prospect Heights, IL: Waveland Press.

MCSHANE, M.D., & WILLIAMS, F.P., III. (1992). Radical victimology: A critique of victim in traditional victimology. *Crime & Delinquency, 38*(2), 258–271.

MORIARTY, L.J., & JERIN, R.A. (1998). *Current issues in victimology research*. Durham, NC: Carolina Academic Press.

PETROVEC, D. (1997). Resurgence of victims. *Social Justice, 24*(1), 163–177.

PHIPPS, A. (1986). Radical criminology and criminal victimization: Proposals for the development of theory and intervention. In R. Matthews & J. Young (Eds.), *Confronting crime* (pp. 189–211). Beverly Hills, CA: Sage.

QUINNEY, R. (1972). Who is the victim? *Criminology, 10*(2), 314–323.

REIMAN, J. (1998). *The rich get richer and the poor get prison*. Boston: Allyn & Bacon.

ROBINSON, M. (1998). Tobacco: The greatest crime in world history? *The Critical Criminologist, 8*(3), 20–22.

ROBINSON, M. (1999). What you don't know can hurt you: Perceptions and misconceptions of harmful behaviors among criminology and criminal justice students. *Western Criminology Review, 2*(1). [On-line]. Available: http://wcr.sonoma.edu/v2nl/v2nl.html.

ROBINSON, M. (in press). The construction and reinforcement of myths of race and crime. *Journal of Contemporary Criminal Justice*.

ROSOFF, S., PONTELL, H., & TILLMAN, R. (1998). *Profit without honor: White-collar crime and the looting of America*. Upper Saddle River, NJ: Prentice-Hall.

RUSH, G. (2000). *The dictionary of criminal justice* (5th ed.) . Long Beach, CA: Dushkin Publishing.

SACCO, V., & KENNEDY, L. (1996). *The criminal event: An introduction to criminology*. Belmont, CA: Wadsworth.

SCHWARTZ, M., & DEKESEREDY, W. (1991). Left realist criminology: Strengths, weaknesses and the feminist critique. *Crime, Law, and Social Change, 15*(1), 51–72.

SCHWENDINGER, H., & SCHWENDINGER, J. (1970). Defenders of order or guardians of human rights? *Issues in Criminology, 7*(1), 72–81.

SGARZI, J.M., & MCDEVITT, J. (2000). *Victimology.* Upper Saddle River, NJ: Prentice-Hall.

SIMON, D., & HAGAN, F. (1999). *White-collar deviance.* Boston: Allyn & Bacon.

TAYLOR, I., WALTON, P., & YOUNG, J. (1972). *The new criminology: For a social theory of deviance.* New York: Harper & Row.

UNITED NATIONS. (1985). *Declaration of basic principles of justice for victims of crime and abuse of power,* adopted at the 96th plenary meeting (November 29, A/RES [29] 40/34).

UNITED STATES DEPARTMENT OF JUSTICE. (1997). *Summary findings* [On-line]. Available at www.ojp.usdoj.gov/bjs/fedle.htm.

UNITED STATES DEPARTMENT OF JUSTICE. (1998). *The number of state and local law enforcement officers assigned to patrol and response duties grew by 19% in four years.* Press release revision, 7/16.

UNITED STATES DEPARTMENT OF JUSTICE. (1999). *Personnel increases in local law enforcement and sheriffs' departments.* Press release, 10/29.

WALLACE, H. (1998). *Victimology: Legal, psychological, and social perspectives.* Boston: Allyn & Bacon.

WILLIAMS, C. (1996). An environmental victimology. *Social Justice, 23*(4), 16–40.

WROBLESKI, H., & HESS, K. (2000). *An introduction to law enforcement and criminal justice* (6th ed.) Belmont, CA: Wadsworth.

ZEHR, H., & UMBREIT, M. (1982). Victim offender reconciliation: An incarceration substitute. *Federal Probation, 46*(1), 63–68.

2

Suitable Responses to Victimization

How Police Should Treat Victims

Amie R. Scheidegger

INTRODUCTION

A criminal justice system that views the state as the injured party in a crime might easily lose sight of the pain and suffering experienced by the individual(s) victimized by an offense. In an effort to protect society as a whole, the needs of the persons most directly impacted by a crime might be overlooked. Thanks to recent legislation and increased awareness of the impact crime has on citizens, the criminal justice system has broadened its definition of the victim and put in place services to assist individuals and communities harmed by crime. The police play a key role in this new victim-centered approach.

How the police respond to victims of crime impacts the outcome of a case, the recovery process of the victim(s), police-community relations, an individual's faith in and respect for the police, and his or her willingness to cooperate with police/law enforcement in the future. It is imperative that police/law enforcement officers are aware of their legal obligations to victims, as well as their humanitarian responsibilities. This chapter will address the question: How should police/law enforcement treat victims? In doing so, one must examine the dynamics of police-victim interactions, how officers assess the needs of a victim, and the importance of training police officers on how to assist crime victims.

POLICE-VICTIM INTERACTIONS

The police are often referred to as the "gatekeepers" of the criminal justice system. They have the power to arrest, issue a verbal or written warning, or take no action at all. If an arrest is made, the formal criminal justice process is enacted. Although an accurate depiction of the police officer's initial role in the criminal justice system, the gatekeeper analogy neglects to incorporate the role the victim of a crime plays in making the police aware of the vast majority of crimes that do not take place in the officer's presence.

It is estimated that over 50% of crimes are never reported to the police (National Victims Center, 1992b). If the police are the "gatekeepers" then the victims are the "messengers" and over half of the messengers never share their information with the gatekeepers. Knowing this, it is critical that police value the victim's effort to report crime and acknowledge the present importance and future implications of police-victim encounters.

Victims cite various reasons for not reporting crimes to the police. Some believe "nothing can be done" and reporting crime is a waste of time for all parties involved. Others feel their victimization is not important enough to warrant reporting to the police. Perhaps the damage to their person or property was insignificant or the value of the item stolen was minimal. Even victims of very serious crimes, such as rape, assault, and domestic violence, might elect not to report their victimization because they fear retaliation, are embarrassed, view the offense as a private matter, or perceive that they caused their victimization in some way. Finally, some individuals do not trust the police and thus do not look toward the police for assistance when they have been harmed.

There is no single reason for not reporting crime to the police. Similarly, there is no one way for police to respond to a crime victim. Future chapters will examine how to interview and respond to various specific types of offenses. Regardless of the offense, police find themselves in a dual role of law enforcer and public servant. Their task is complex. They have a duty to collect evidence in an attempt to solve a crime and also the responsibility to provide services to the victim, which might enable the victim to recover more fully from his or her victimization. A balance between these two functions must be made if both goals are to be achieved.

Federal and state legislation dictates the legal duties of an officer and the rights of the victim. Within this framework, departmental policies also dictate specific procedures officers must follow when responding to calls for service. An officer must adhere to both departmental policies and legal responsibilities to best serve the victim.

In its report, *New Directions from the Field: Victims' Rights and Services for the 21st Century* (1998), the Office for Victims of Crime (OVC) published several recommendations for police/law enforcement officers on how to respond to victims. The report included policies, protocols, and procedures for a comprehensive police/law enforcement response to victims of crime. The general recommendations of the report included the following: police/law enforcement providing victims with verbal and written notification of their rights, utilizing community partnerships to ensure victims have access to needed services and programs, and notifying victims of the status of the case. Table 2-1 contains a full list of the recommendations made by the OVC.

Being the victim of a crime is difficult enough without experiencing a "second victimization"—that is, insensitive treatment at the hands of the criminal justice system.

TABLE 2-1   How to Respond to Crime Victims

1. Upon first contact with law enforcement, the responding officer should give victims verbal and written notification of their rights according to state or federal law.
   - This can be accomplished by giving victims a card that specifies their rights in accordance with state or federal law, often referred to as a "Reverse Miranda" card.
   - It is imperative that such information be language and age appropriate. Brochures on emergency and crisis services, and crime victims' compensation should be developed in different languages—as well as for victims with physical and/or mental disabilities—and distributed appropriately.

2. Law enforcement agencies should utilize community partnerships to ensure that victims have access to the following emergency services, financial assistance, information, and community programs:
   - On-site crisis intervention, assistance, and support, either by a trained law enforcement officer or through on-site support from a victim services professional.
   - Immediate referrals, verbally and in writing, to community agencies that offer emergency services, emergency financial assistance, 24-hour crisis intervention, shelter, and transportation. Proper referrals should include current names and telephone numbers of private and public victim assistance programs that provide counseling, treatment, and other support services.
   - Transportation and accompaniment to emergency medical services if they are injured.
   - A brochure or other written resources that explain the expected reaction victims have to specific crimes.
   - Written information about crime victim compensation and how to apply for it.
   - Victims should not be charged for certain medical procedures or for costs arising out of the need to collect and secure evidence.

3. Protection from intimidation and harm:
   - Verbal and written notification about the procedures and resources available for the victim's protection.
   - An explanation of antistalking rights, availability of emergency protection orders, other protection from intimidation and harassment measures, as well as information on victim safety and security.
   - Victim notification of the release of the accused and inclusion of "no contact with the victim" orders as conditions of the release.

4. Investigation:
   - A verbal and written orientation to the investigation process.
   - Procedures allowing victims to choose an individual to accompany them to interviews.
   - The name and telephone number of the law enforcement officer investigating the offense and the arrest, and the police report number or any other identifying case information.
   - A free copy of the incident and arrest report.

(continued)

TABLE 2-1   *Continued*

5. If an arrest has been made, victims should be notified of:
   - The arrest of the defendant.
   - The next regularly scheduled date, time, and place for initial appearance.
   - Any pretrial release of the defendant.
   - Their rights within the criminal and juvenile justice processes, including the right to be present at all justice proceedings that the accused, defendant, and/or prisoner has the right to attend, and the right to be heard, both orally and/or in writing, at various stages of the case.
   - Upon release of the suspected offender, notification of the date, time, and place of the next court appearance, and how to obtain additional information about the subsequent criminal proceedings.
6. If there is no arrest within 7 days:
   - Information about the right to notification of arrest, providing the victim maintains a current address, regardless of the length of time between the commission of the crime and date of arrest.
7. If the case has been submitted to a prosecuting attorney's office:
   - Give notification of the name, address, and telephone number of the prosecuting attorney assigned to the case.
8. Prompt property return:
   - Speedy return of property held by law enforcement with victims provided with verbal and written information on how to obtain their property.
   - Free storage of the victim's property.
   - Reimbursement for the actual replacement costs of any property that is lost, sold, or damaged while being held as evidence.

Source: Adapted from *New Directions from the Field: Victims' Rights and Services for the 21st Century,* Office for Victims of Crime, U.S. Department of Justice, 1998.

When officers ask victims insensitive questions, insinuate or suggest victims contributed to their victimization, neglect to return property held as evidence, or are in any way disrespectful toward victims, the result is a second victimization. Being aware of the needs of the victim, the consequences of crime for the victim, the common responses to victimization (see Chapter 8), and the additional needs of distinct victim populations, officers can greatly diminish or eliminate their part in the second victimization process. Adhering to the aforementioned policies, protocols, and procedures recommended by the U.S. Department of Justice is another way to avoid second victimization.

VICTIMS' NEEDS

The needs of crime victims are as diverse as the individuals who are victimized. To construct a grocery list of needs for all victims would be impossible. However, victims generally want information, recognition, advice, support, protection, and reassurance. The majority of citizens, in general, are ill informed about the duties and responsibilities of the police. They are unfamiliar with the legal system and even less aware of their own personal rights. Victims are no exception.

Victims are thrust into a system that is unfamiliar and often are forced to interact with police/law enforcement personnel whom they may distrust. It is only natural for victims to want and need information from the police. The police must provide victims with various types of information during the initial contact. The legal process, rights of the victims, and services available to assist victims are a few of the topics police officers must discuss with the victim.

Victims also want their injuries to be recognized and to be taken seriously (Skogan, 1989). Careful collection of evidence and effort taken at the crime scene not only aids in solving a crime but also in illustrating to the victim the officer is taking the victimization seriously. Victims tend to rate officers who express caring and supportive attitudes more favorably, factoring into their assessment the amount of time and trouble the officer took handling their case (Maguire, 1982; Shapland, 1984).

In addition, victims need reassurance and protection. Feelings of isolation and vulnerability are common after victimization. Some victims express the fear no one will come to their aid if need be. Police response time can help to foster or diminish that fear. How quickly police respond to a call for assistance can influence one's perception of police professionalism and their ability to protect the victim (Jerin & Moriarty, 1998). Percy (1980) found victim satisfaction with police depended on how closely expected arrival time matched actual arrival time.

Unrealistic expectations that victims sometimes have of policing/law enforcement are problematic when one is attempting to assist a victim. Crime victims learn about policing and law enforcement from media sources (television, movies, and news primarily) that often portray policing/law enforcement in unrealistic or negative ways, making crime victims unsure of what to expect or giving them unrealistic expectations of the services law enforcement can provide (T. Clawson, personal communication, December 3, 1999). The typical crime drama on television portrays officers as able to solve any crime, no matter how difficult or how much evidence is lacking. The "competent" officer always catches the criminal and it only takes one hour.

A second issue that influences victims' expectations of policing/law enforcement is prior experiences. Some victims have had positive experiences with the police in the past whereas others have had negative experiences. Victims expect similar (positive or negative) treatment in the future and base their expectations of the police/law enforcement accordingly. One key component to changing negative expectations of the police is to attempt to make all police-victim interactions as positive as possible (T. Clawson, personal communication, 1999).

Because crime victims are typically unfamiliar with how policing and law enforcement operate and sometimes anticipate encounters with officers to be negative experiences, the officer must carefully and fully inform victims of their legal rights, the investigation process, and court proceedings. Providing victims with this vital information, verbally and in writing, can help to make the experience more positive for the victim and, to some extent, more positive for the officer as well. Uncertainty and confusion can lead to frustration and misunderstandings which, in turn, result in negative expectations, lower satisfaction, and less cooperation from citizens.

CONSEQUENCES OF CRIME

The seriousness of a crime and presence of injuries appear to be two of the most important factors police utilize when deciding how to respond to a crime (Cole, 1995). Obviously, physical injuries in need of immediate medical attention are of particular concern to officers at a crime scene. However, injuries take many different forms and the police must be able to recognize and appreciate both the short- and long-term consequences of victimization.

Depending on the type of offense, victims might suffer physical, psychological, and/or financial consequences. Recognizing these types of damages will allow the officer to better serve the needs of the victim. Although police officers are not expected to be physicians, they should have a basic understanding of the various types of injuries victims might suffer (Wallace, 1998). Some physical injuries are more obvious than others. Visible wounds, such as contusions, lacerations, bullet wounds, severe broken bones that penetrate the flesh, and burns, warrant emergency medical attention and are considered to be immediate physical consequences of victimization.

The long-term physical consequences of injuries, such as scars and paralysis, will affect the victim long after the police have gone and the case has been closed. For some victims, another long-term physical effect of victimization may be the contraction of a disease that will lead to inevitable death or long-term illness. Sexually transmitted diseases (STDs), such as HIV/AIDS, gonorrhea, and herpes, along with nonsexually transmitted diseases, such as hepatitis, pose serious long-term effects on the victim. The officer must be familiar with the ways by which such diseases are contracted to ensure the victim receives proper medical attention as soon as possible.

Just as physical injuries resulting from crime differ by offense type, so too do the financial injuries. The financial consequences of crime are not limited to the tangible losses of money taken or the value of property damaged or stolen. Financial losses also include intangible losses such as lowered quality of life, emotional pain and suffering, fear, and distrust. Even losses with a specific monetary value affect victims differently. Medical bills cause different financial costs to victims with medical insurance than to those without medical insurance. Insured property owners might have very different financial losses when compared to the uninsured victim. Victims with low socioeconomic status might suffer differently than victims with higher socioeconomic status. Police officers must be conscious of the different impact even similar offenses have on different people.

In 1996, Miller, Cohen, and Wiersema published a report estimating the average costs of various crimes. Tangible costs included in the calculation were medical care, mental health care, police and fire services, victim services, and productivity. The tangible cost "productivity" warrants clarification. Included within this category were the costs associated with lost wages, benefits, housework, and missed school days resulting from victimization. Secondary victims, such as family members and friends, might suffer similar productivity losses when caring for the victim, accompanying them to court, or taking them for medical treatment. Coworker productivity is affected by victimization as well. Increased workload or an inability to perform one's job as a result of the victim's absence from work impacts coworkers' productivity. Accommodating and retraining employees permanently or temporarily disabled as a result of victimization can impact employer productivity. The report concluded that the average tangible costs of a murder are $1,030,000; rape/sexual assault, $5,100; robbery and attempted robbery with injury, $5,200; assault and attempted assault, $1,500; and burglary or attempted burglary, $1,100.

Victimization is not limited to tangible costs, however. More difficult to quantify, yet equally damaging, are the intangible costs suffered by the victim. Intangible costs include pain, suffering, fear, and reduced quality of life. The emotional trauma that victims suffer can be long lasting. The burglary victim who no longer feels safe in his or her home, the loss of a loved one at the hands of an offender, and the nightmares suffered and sleep lost as a result of crime all constitute intangible costs. Both primary and secondary victims suffer intangible losses which dramatically increase the total cost of crime. For example, the intangible costs of murder ($1,910,000) in addition to the tangible costs ($1,030,000) result in a total cost of $2,940,000. The intangible costs of rape and sexual assault are estimated to be $81,400, making the total cost $86,500. Assault and attempted assault have an intangible cost of $7,800, which is significantly greater than the $1,500 tangible costs (Miller, Cohen, & Wiersema, 1996).

Police officers must be aware of both the tangible and intangible costs of victimization. Although costs differ from one victim to another, the totality of victimization must be in the forefront of the officer's mind when assisting the victim. Items of little monetary value, (stolen in a burglary, for example) might have irreplaceable sentimental value to the victim. To assume the victim is overreacting or responding irrationally to the loss of inexpensive property is to negate the intangible costs of crime altogether and to trivialize the individual's status as a victim.

UNDERSTANDING RESPONSES TO VICTIMIZATION

There is no way of guaranteeing how a victim will respond to crime. The immediate medical needs of the victim, tangible and intangible costs to the victim, and type of offense all impact the victim's response to crime. The shock of becoming a victim is often followed by feelings of fear, anger, shame, self-blame, helplessness, and depression (Finn & Lee, 1987). These responses contribute to the immediate stress experienced by the victim and even long-term psychological effects. Police officers must have a general knowledge of typical responses to victimization to understand the emotions the individual might be experiencing and to best serve the victim. Also, officers must keep in mind that victims vary in their response to crime and thus should not be judgmental or draw conclusions about the victim based solely on behaviors exhibited at the scene of a crime.

Experiencing a traumatic event, such as crime, results in emotional, psychological, and physiological responses. The traumatic event and its effects are also referred to as a crisis. Crisis can be defined as a specific set of temporary circumstances that result in a state of upset and disequilibrium, characterized by an individual's inability to cope with a particular situation using customary methods of problem solving (Bard & Sangrey, 1986; Roberts, 1990). In his book, *Victimology: Legal, Psychological, and Social Perspectives,* Harvey Wallace (1998) describes a three-stage process of crisis reaction.

The first stage is the impact stage. Immediately following the crime, victims express feelings of disbelief that the crime occurred, a sense of shock, and physical reactions such as loss of appetite and the inability to sleep. Increased feelings of vulnerability and helplessness are also common at this stage.

The length of time a victim experiences the impact stage varies from minutes to hours, or even days. It is very likely that police officers, as the initial contact persons, will encounter the victim at this stage of the crisis process. Mood swings are not uncommon for victims during this stage. Victims, in a state of emotional vulnerability, might misinterpret comments and reactions of those around them, including friends, family members, and the police. Other victims might display "flat affect," which is characterized by a general lack of emotional responses (T. Clawson, personal communication, 1999). It is crucial that police officers are aware of the possible heightened sensitivity and emotional reactions of victims early in the crisis process.

The second stage in the crisis process is the recoil stage. A wide variety of responses are common during this stage. Victims may experience guilt, anger, self-pity, grief, and denial. Victims may feel guilty for not doing more to prevent their victimization or for bringing shame, embarrassment, or pain to their loved ones. Anger toward the offender is also common during this stage. Some victims will express rage or violent behavior. Others might fantasize about getting revenge on the perpetrator. Many victims expect the police, the court system, and the correctional system to achieve that revenge. Officers should remember that the need for revenge is a natural part of the healing process, but not always a response to victimization.

Denying the crime occurred is also a common response. This response may also account for many unreported crimes. In the case of particularly violent or personal crimes, the victim might not be equipped to deal with the events that took place. Some victims will repress the events, rationalizing the crime as a bad dream. Some will project the crime, referring to the events as if they occurred to someone else. Denial allows victims to remove themselves emotionally from the situation, often in an attempt to carry on with their lives as if the incident never happened.

The third stage, according to Wallace, is the reorganization stage. As feelings of rage, anger, fear, and denial diminish, victims are able to begin to put their experiences into perspective. Although the experience is not forgotten, normal daily activities and concerns begin to reenter victims' lives.

In addition to being aware of the common stages of the crisis process, officers must be familiar with long-term emotional and psychological disorders that can result from victimization. Developing a psychological disorder, such as post-traumatic stress disorder (PTSD), acute stress disorder (ASD), or depression is not uncommon for crime victims. In a 1992 study of PTSD in rape victims, 94% of the victims exhibited symptoms of the disorder one week after the crime. Twelve weeks after the assault, 47% continued to exhibit symptoms (Rothbaum, Foa, Murdock, Riggs, & Walsh, 1992). PTSD is not limited to victims of sexual assault. Any traumatic event, criminal or otherwise, can result in PTSD. The following is a brief, introductory discussion of PTSD including the characteristics of the disorder and the possible impacts PTSD can have on crime victims. (A more lengthy discussion of PTSD is found in Chapter 4 of this text.)

POST-TRAUMATIC STRESS DISORDER

Post-traumatic stress disorder is defined in the American Psychiatric Association *Diagnostic and Statistical Manual of Mental Disorders* IV (1994, p. 427) as the development of characteristic symptoms following exposure to an extreme traumatic stressor involving direct personal experience of an event that involves actual or threatened death or serious injury, or other threat to one's physical integrity. Symptoms of PTSD are widespread among victims of crime from burglary victims to attempted homicide survivors. Characteristics of PTSD are listed in Table 2-2.

TABLE 2-2   Characteristics of Post-Traumatic Stress Disorder

A. The person has been exposed to a traumatic event in which both of the following were present:
  - The person experienced, witnessed, or was confronted with an event or events that involved actual or threatened death or serious injury, or a threat to the physical integrity of self or others.
  - The person's response involved intense fear, helplessness, or horror.

B. The traumatic event is persistently reexperienced in one (or more) of the following ways:
  - Recurrent and intrusive distressing recollections of the event.
  - Recurrent distressing dreams of the event.
  - Acting or feeling as if the traumatic event were recurring: reliving the experience, illusions, hallucinations, and dissociative flashback episodes.
  - Intense psychological distress to cues that symbolize or resemble an aspect of the traumatic event.
  - Physiological reactivity on exposure to cues that symbolize or resemble an aspect of the traumatic event.

C. Persistent avoidance of stimuli associated with the trauma and numbing of general responsiveness (not present before the trauma), as indicated by three (or more) of the following:
  - Efforts to avoid thoughts, feelings, or conversations associated with the trauma.
  - Efforts to avoid activities, places, or people that arouse recollections of the trauma.
  - Inability to recall an important aspect of the trauma.
  - Markedly diminished interest or participation in significant activities.
  - Feelings of detachment or estrangement from others.
  - Restricted range of affect (e.g., unable to have loving feelings).
  - Sense of a foreshortened future (e.g., does not expect to have a career, marriage, children, or a normal life span).

D. Persistent symptoms of increased arousal (not present before the trauma), as indicated by two (or more) of the following:
  - Difficulty falling or staying asleep.
  - Irritability or outbursts of anger.
  - Difficulty concentrating.
  - Hyper-vigilance.
  - Exaggerated startle response.

Source: Reprinted with permission from the *Diagnostic & Statistical Manual of Mental Disorders, Fourth Edition.* Copyright 1994 American Psychiatric Association.

The impact this disorder has on the individual's quality of life and range of behaviors is dramatic. PTSD sufferers are repeatedly violated by their experience. Living the event over and over again in their mind, detaching themselves from friends and family, losing aspirations and interests, and increasing their consumption of alcohol and drugs are only a few of the responses brought on by PTSD. In addition, poor impulse control can result from PTSD, and sufferers run an increased risk of harming themselves and others (Meadows, 1998). Victims may attempt suicide or resort to committing crimes as a result of their victimization. Furthermore, because not all victims will experience or suffer from PTSD or respond to being a victim in the same manner, one should be aware of the distinction among victims by population types.

Distinct Victim Populations

The type of crime a person is the victim of has a significant impact on the needs of the victim and the officer's response. Officers should be aware of personal characteristics of the victim that might also impact the needs of the victim. Personal characteristics such as age, gender, ethnicity, and mental/physical abilities must be taken into consideration when responding to crime victims. A few such personal characteristics by victim population type are offered.

Elderly Victims

Elderly citizens fall prey to the same range of offenses as younger citizens. Elderly abuse, that is, conduct that results in the physical, psychological, or material harm through neglect or injury, can occur in domestic and institutional settings (Meadows, 1998). Elderly victims are physically assaulted by family members, caregivers, and strangers. They are psychologically abused, emotionally manipulated, and financially exploited (Bachman, 1992). The elderly are the targets of fraud more often than any other age group, sometimes losing their entire life savings (Boles & Patterson, 1997). The status of the elderly in society may further exacerbate the victimization problem through increased isolation and their removal from society's watchful eye as many elderly are homebound or live in extended care facilities (Pillemer & Suitor, 1988).

The impact of victimization can be vastly different for the elderly victim than the younger victim. For example, the physical condition and health of the elderly victim can create additional medical needs. As bones become more brittle with age, less force or trauma is needed to break them. Bruising occurs more easily and severely, and daily medications may interfere with the long-term healing process.

Financially, the costs of victimization may also be greater for the elderly victim. Victims on a fixed income might lack the resources needed to replace stolen and damaged property, and additional medical attention can be financially devastating to the elderly. Officers must be prepared to provide information on social services and emergency financial assistance programs to elderly victims.

The emotional needs of the victim may also differ in intensity from those of younger victims. Sentimental attachments to belongings can be far greater for the elderly. Fear, helplessness, and vulnerability might be felt more immediately by these victims. Additionally, elderly victims may be even less likely to share information they feel is personal or embarrassing (Wallace, 1998). Police officers might experience a "generation gap" whereby members of older generations consider some topics off-limits, such as sex or personal finances, regardless of the circumstances. The result of increased awareness of the needs of elderly victims has led to states enacting specific legislation regarding elderly victims, and social service organizations exist in many communities devoted solely to assisting elderly individuals (Patterson & Boles, 1992).

Victims with Disabilities

Officers responding to physically disabled victims must be accommodating to the special needs of the victim. Elderly victims may also exhibit many of the same physical disabilities included in this group of victims. The National Victims Center (1992a, p. C-13) published "Eight Do's and Don'ts in Working with Disabled Victims of Crime" which should be helpful to officers when assisting disabled victims. They are:

- Look directly at the victim when speaking. Deliberately averting your eyes is impolite and can be uncomfortable.

- Feel free to ask a disabled victim how you should act or communicate most effectively if you have any doubts about correctness in the situation.

- Address and speak directly to the disabled person, even if the person is accompanied or assisted by a third-party nondisabled person.

- Feel free to offer physical assistance to a disabled person, such as offering your arm if the need arises, but do not assume the person will need it or accept it.

- Ask a disabled person about any personal needs that will require special services or arrangements, and then attempt to make arrangements to meet those needs.

- Do not stare or avoid looking at a visible disability or deformity or express sympathy to the disabled victim

- Do not tell the disabled victim you admire his or her courage or determination for living with the disability. The disabled person doesn't want to be thought of as a hero.

- Do not avoid humorous situations that occur as a result of a disability. Take your cue from the victim.

Hearing-impaired (individuals who have some ability to hear) and deaf (individuals who cannot hear at all) victims also require special accommodations from the police. Verbal communication may be difficult for both the victim and the officer. The officer should begin by asking the victim, verbally or by writing a note, how he or she would like to communicate. This simple act can alleviate some of the frustration that can result from communication problems. According to Boles and Patterson (1997), the question should be phrased: "Would you feel more comfortable using American Sign Language and having an interpreter?" If the victim prefers sign language, every effort should be made to accommodate the victim. However, an officer or victim advocate trained in sign language might not be available to assist the officer responding to a call. In the event this occurs, nontraditional forms of communication might be necessary.

Telecommunication devices for the deaf (TDD) can be very helpful in making a report or investigating a crime. Many large police departments have TDD terminals within the dispatch department that should be utilized if available. Other departments rely on various community service programs that relay calls from TDD terminals. Whichever the case, the officer must be cautious that information is not relayed improperly or left out when using a third party to gather information. Other nontraditional forms of communication such as written communication, gestures, and drawings may be helpful in assisting the hearing-impaired or deaf victim. Goddard (1989) suggests utilizing anatomically correct dolls to assist young sexual assault victims and those with communication limitations to identify the nature, details, and circumstances of the assault.

Some officers might assume they can ask family members to interpret for the hearing-impaired or deaf victim. This is not an appropriate role for the family member to carry out. The officer must keep in mind the very real possibilities that (1) the family member might be the assailant, (2) the victim might be uncomfortable discussing sensitive details in front of the family member, and (3) the family member might also be a victim in the case.

Visually impaired and blind victims also require special accommodations. The term *visually impaired* indicates that there is limited sight; the term *blind* is a legal term indicating severe loss of sight. The police officer responding to a visually impaired or blind victim must be prepared to provide the victim with printed information in an alternative written format, such as Braille or large print. Large print information can be easily generated using a word processor program that offers large font options. In addition, large print may also be useful for elderly crime victims whose eyesight may be diminished. Audiotapes are also an acceptable means of sharing information with the visually impaired or blind victim. Illiterate victims can also benefit from documentation provided in audio format.

Officers should not make assumptions about a victim's ability to identify an unknown assailant merely because he or she did not see the perpetrator. The victim might be able to identify the assailant's voice or some other characteristic. For example, a victim in a sexual assault case was able to identify her attacker by his body odor (Chicago Police Department, 1990). Expert witnesses can testify that many blind and visually impaired individuals are able to compensate for their loss of vision by increasing their abilities in other senses (1990). Remember, asking thorough yet sensitive questions can be crucial to the outcome of the case, victim satisfaction with the police, and the recovery process for the victim.

Developmentally disabled victims are an additional group with special needs and considerations. The term *developmentally disabled* includes "mental retardation, cerebral palsy, or other conditions that might interfere with normal (physical and/or mental) development and cause difficulties in communication, thinking, and mobility" (Baldarian, 1992, p. 8). The officer responding to a developmentally disabled victim should not assume the victim is incompetent as a witness. Though the comprehension and communication levels of the victim must be determined, developmentally disabled victims can offer pertinent evidence and testimony. Care must be taken when questioning the victim to ensure the officer does not misinterpret the victim or mislead the victim in any way which may negatively impact the case. Using language appropriate for the mental age of the victim is crucial (Boles & Patterson, 1997). Also, victims must be allowed to answer questions using phrases and terminology (slang) that are familiar to them. The caregiver should be asked for assistance only after he or she has been ruled out as a suspect and only with the victim's permission (Chicago Police Department, 1990).

Ethnically Diverse Victims

Police officers in a multicultural society must be educated about issues of diversity. As public servants, police officers have a responsibility to know the public they serve. Armed with an appreciation for diversity and an understanding of the possible needs associated with different ethnic groups, the officer will be better prepared to respond to all crime victims. It is beyond the scope of this chapter to discuss all of the various possible needs of each ethnic group in our society. Thus, this discussion will be limited to two general considerations officers should be conscious of when assisting ethnically diverse victims: communication needs and cultural needs.

Different communication styles can be problematic during police-victim interactions. Familiarity with the customs and traditions of various ethnic groups can facilitate better communication between the victim and officer. For example, direct eye contact is considered disrespectful in Asian and American Indian cultures (Rasche, 1995). The officer could offend the victim through the use of direct eye contact or misinterpret the victim's avoidance of eye contact as deceit or dishonesty. Some ethnic groups are more vocal and animated when communicating than others. The victim who is talking loudly and using a lot of hand gestures should not automatically be interpreted as "out of control" or a "threat to the officer's safety." These forms of expression could be acceptable in the victim's culture.

Obviously, an additional communication need could be the use of an interpreter. Whether the victim is a permanent resident, vacationer, or illegal immigrant, the victim must be given all of the rights and considerations afforded by law. All written materials routinely given to victims should be available in the language necessary to inform victims of their rights and about the legal process.

The officer and the victim must be able to communicate in some way if services are to be rendered. As with other victim groups for which communication can be difficult, it is not appropriate to ask family members to serve as interpreters in cases where the family member could be the assailant. Also, keep in mind that sensitive information can be embarrassing for the victim to share with family members. A useful tool when assisting non-English-speaking victims is a pocket translator specifically written for police/law enforcement officers. In their book, *Pocket Partner* (1999), Evers, Miller, and Glover have included questions and phrases commonly used in police work written in both English and Spanish including phonetic pronunciations.

A second consideration officers must make when responding to ethnically diverse victims is the impact crime has on the victim. Within some cultures particular crimes may have considerably different negative impacts, and greater stigma can be attached to the crime victim by family members and the community. In the Asian community, for example, the fear of being ostracized or shamed as a result of a sexual assault is so great that women often do not report the crime (Boles & Patterson, 1997). Hispanic women often find speaking to men about sexual assault particularly uncomfortable. Many illegal immigrant victims fear that reporting crime will result in deportation, and they also lack supportive family ties to help with the recovery process (Rasche, 1995).

There is a myriad of reasons ethnic victims need and deserve special considerations from the police. Some reasons stem from communication difficulties. Other factors are based on cultural differences. The range of acceptable and unacceptable behaviors within an ethnic group can be quite different from those of the dominant culture. Officers who are

---

TABLE 2-3   Acceptable Death Notification Practices

Death notification practices:

- Be absolutely certain of identity.
- Go to the residence. Do not call.
- Make notification in a timely manner.
- Notify victim advocate as soon as possible.
- Take someone with you/two cars if possible.
- Present credentials; ask to come in.
- Sit down. Ask them to sit down. Be sure you have nearest next of kin.
- Inform simply and directly (with compassion).
- Answer all questions honestly.
- Provide written information.
- Provide transportation if identification of the body is necessary.

What to say:

- I'm so sorry.
- It's harder than most people think.
- Most people who have gone through this react similarly to what you are experiencing.
- Is there anything you would like to tell me or ask me?
- I'll check back with you tomorrow, to see how you are doing and if there is anything more I can do for you and your family.

Source: South Carolina Department of Public Safety, 1999.

aware of differences, sensitive to the needs of the ethnic victim, and fully prepared to deal with different groups of victims will inevitably respond more effectively.

SECONDARY VICTIMS

As mentioned throughout this chapter, in addition to the primary victim of a crime, there are also secondary victims. The friends, family members, coworkers, and neighbors of the victim all constitute secondary victims. Victim's rights legislation was written primarily in reference to the primary crime victim. Police officers nonetheless have a responsibility to secondary victims, both as citizens of a community and as persons in need of service, regardless of the lack of a legal duty. Secondary victims should have access to social services, support groups, and counseling services (Patterson & Boles, 1992). Secondary victims should be given written information about the services available to them similar to the information given to primary victims. Officers must be cautious not to share confidential information with a secondary victim without the permission of the primary victim. It is still the rights of primary victims that are paramount in every case.

TABLE 2-4 Unacceptable Death Notification Practices

What NOT to say:

Discounting and patronizing statements:

- I know how you feel.
- Time heals all wounds.
- You'll get over this.
- You must go on with your life.
- He/she didn't know what hit them
- You must focus on your precious memories.
- You can't bring them back.
- God clichés.
- At least you still have your other children.

Disempowering statements:

- You don't need to know that.
- I can't tell you that.
- What you don't know won't hurt you.

Source: South Carolina Department of Public Safety, 1999.

One of the most difficult aspects of a police officer's job is notifying surviving family members about the death of a loved one. The notification of a loved one's death will remain in the survivor's memory forever. The manner in which the notification is made is of the utmost importance. Police officers do not only serve death notifications for individuals killed as a result of crime; notifications are also made for deaths resulting from natural causes or accidents. Table 2-3 contains a list of death notification practices, and acceptable things to say to secondary victims and next of kin. Table 2-4 contains unacceptable death notification practices.

TRAINING

The training of police/law enforcement in how to respond to victims has progressed significantly in recent years. Training in victims' rights and needs results in increased compassion, comprehensiveness, and professionalism in police/law enforcement officers. Special units within departments devoted to investigating domestic violence, sexual assault, child abuse, and elderly abuse can be beneficial in assessing the needs of a victim and assisting the victim in finding help within the community. However, all officers must be able to respond appropriately to the needs of various types of victims and to various types of crimes.

In the aforementioned report, *New Directions from the Field: Victim's Rights and Services for the 21st Century,* recommendations in regard to training include that "all law enforcement personnel, from dispatchers through management, should receive initial and on-going training about the impact of crime and how to respond sensitively and effectively to victims" (p. 64). The report called for additional instruction in the areas of interviewing victims, responding to victims, and victims' rights training. In his 1989 study of crime victim interviewing techniques, MacLeod reported an increase in both the type and amount of information given to police officers who took the nature of the offense into consideration when interviewing assault victims. As discussed in this chapter, the police must have a keen awareness of the various impacts crime has on victims. The officer must be well trained to understand how crime impacts the victim physically, psychologically, emotionally, and financially. An officer must understand the crisis process as it applies to victims and how to assist special victim populations. The more thoroughly trained an officer is, the more likely he or she will be able to respond adequately and appropriately to crime victims.

CONCLUSION

Victims of crime serve a vital role in the criminal justice system. Each victim can provide police/law enforcement with information that is necessary to solve a crime. It is imperative that officers are aware of how to best ascertain the needed information while remaining sensitive to the needs of victims. One essential component of policing is interviewing the victim. How the interview is conducted may determine the quality and quantity of information gathered. Officers must be aware of the financial, psychological, and physical impacts crime has on victims. To properly serve the community and its citizens each department must make a commitment to protect and serve those most in need of protection and service—victims. Different groups of victims, such as the elderly, physically or mentally disabled, and ethnically diverse populations, may pose unique challenges to police/law enforcement. However, all victims must be afforded all of the rights and protections guaranteed to them under the law. Providing information about victims' rights, the court process, and services that are available in the community is only one small way the officer can assist victims. In order to provide such information, the officer must be prepared at all times with documents written in alternative formats, contact numbers for service providers, and a willingness to extend a helping hand to victims.

REFERENCES

AMERICAN PSYCHIATRIC ASSOCIATION. (1994). *Diagnostic and statistical manual of mental disorders* (4th ed.). Washington, DC: Author.

BARD, M., & SANGREY, D. (1986). *The crime victim's book* (2nd ed.). New York: Brunner/Mazel.

BACHMAN, R. (1992). Elderly victims. *Special Report, Bureau of Justice Statistics.* Washington, DC: U.S. Department of Justice.

BALDARIAN, N.J. (1992). RAPPORT model aids victims with developmental disabilities. *NRCCSA News, 1*(4), 8.

BOLES, A.B., & PATTERSON, J. (1997). *Improving community response to crime victims: An eight-step model for developing protocol.* Thousand Oaks, CA: Sage.

CHICAGO POLICE DEPARTMENT. (1990). *Detective division protocol for sex crimes investigations.* Unpublished manuscript.

COLE, C.F. (1995). *The American system of criminal justice.* Belmont, CA: Brooks/Cole.

EVERS, D., MILLER, M., & GLOVER, T. (1999). *Pocket partner.* Littleton, CO: Blue Willow.

FINN, P., & LEE, B. (1987). *Serving crime victims and witnesses.* Washington, DC: National Institute of Justice.

GODDARD, M.A. (1989). *Sexual assault: A hospital/community protocol for forensic and medical examination.* Rockville, MD: National Criminal Justice Reference Service.

JERIN, R.A., & MORIARTY, L.J. (1998). *Victims of crime.* Chicago: Nelson-Hall.

MACLEOD, M. (1989). Interviewing victims of crime. In E. Viano (Ed.), *Crime and its victims: International research and public policy issues.* Washington, DC: Hemisphere.

MAGUIRE, M. (1982). *Burglary in a dwelling.* London: Heinemann.

MEADOWS, R. (1998). *Understanding violence and victimization.* Upper Saddle River, NJ: Prentice-Hall.

MILLER, T., COHEN, M., & WIERSEMA, B. (1996). *Victim costs and consequences: A new look.* National Institute of Justice. Washington, DC: U.S. Department of Justice.

NATIONAL VICTIMS CENTER. (1992a). Eight do's and don'ts in working with disabled victims of crime. *Focus on the future: A systems approach to prosecution and victim assistance.* Washington, DC: U.S. Department of Justice.

NATIONAL VICTIMS CENTER. (1992b). *National crime victims survey.* Washington, DC: U.S. Department of Justice.

OFFICE FOR VICTIMS OF CRIME. (1998). *New directions from the field: Victims' rights and services for the 21st century.* Washington, DC: U.S. Department of Justice.

PATTERSON, J., & BOLES, A.B. (1992). *Looking back, moving forward: A guidebook for community responding to sexual assault.* Arlington, VA: National Victim Center.

PERCY, S. (1980). Response time and citizen evaluation of police. *Journal of Police Science and Administration, 8,* 75–86.

PILLEMER, K., & SUITOR, J. (1988). Elder abuse. In V.B. Van Hasselt, et al. (Eds.), *Handbook of family violence.* New York: Plenum.

RASCHE, C. (1995). Minority women and domestic violence: The unique dilemmas of battered women of color. In B. Price & N. Sokoloff (Eds.), *The criminal justice system and women: Offenders, victims, and workers* (2nd ed., pp. 246–261). New York: McGraw-Hill.

ROBERTS, A. (1990). *Crisis intervention handbook: Assessments, treatment and research.* Belmont, CA: Wadsworth.

ROTHBAUM, B., FOA, E., MURDOCK, T., RIGGS, D., & WALSH, W. (1992). A prospective examination of post-traumatic stress disorder in rape victims. *Journal of Traumatic Stress, 5,* 455–475.

SHAPLAND, J. (1984). Victims, the criminal justice system, and compensation. *British Journal of Criminology, 24,* 131–149.

SKOGAN, W.G. (1989). The impact of police on victims. In E. Viano (Ed.), *Crime and its victims: International research and public policy issues* (pp. 71–76). Washington, DC: Hemisphere.

SOUTH CAROLINA DEPARTMENT OF PUBLIC SAFETY. (1999). *South Carolina Criminal Justice Academy training manual.* Columbia, SC: Author.

WALLACE, H. (1998). *Victimology: Legal, psychological, and social perspectives.* Boston: Allyn and Bacon.

3

Victim Reporting

Strategies to Increase Reporting

Peter J. Mercier

INTRODUCTION

The police do not discover most crimes. Rather, the majority of crimes are reported to them (Reiss, 1971). Early victimization surveys indicate that only 3% of those who reported having been victimized received the attention of law enforcement because the police were already on the scene (Hindelang & Gottfredson, 1976). Such research suggests that whether police even learn of a crime is largely determined by what the victim decides to do (Greenberg & Ruback, 1992)—report or not report.

In most instances, it is the victim who initiates the criminal justice process. Having the choice to withhold or provide information regarding the commission of a crime against them, victims heavily influence the mission of the criminal justice system. Just as their willingness to report crimes can assist in bringing a guilty perpetrator to justice, their hesitancy or refusal to report crimes prohibits the police from doing their jobs. Without the cooperation of victims, the system may collapse.

When a victim *does* report a crime to the police, the case *may* proceed through the system, although this is by no means certain. The criminal justice system is a filtering process; there is no guarantee that a reported crime will lead to an arrest or that an arrested perpetrator will be adjudicated. Prosecutors sometimes drop cases brought to them by police. There may be insufficient evidence, or the evidence may be viewed as weak. If the likelihood of winning a case is perceived as low, the prosecutor may be reluctant to press charges. Perhaps the accused has a reasonable alibi, or the case lacks probable cause. In addition to a lack of evidence, a case may be dismissed at preliminary hearings due to a violation of the suspect's constitutional rights. Even if a case does go to trial, the defendant may be acquitted.

However one views the criminal justice system, whether as complex legal maneuvering or the protection of inherent rights, the fact remains that without victim reports of crime to the police, there exists little chance that suspects will be arrested or brought to trial (Hindelang & Gottfredson, 1976). Playing a critical role in the operation of the criminal justice system, victims impact the police's decision to make arrests, the prosecutor's decision to file charges, and the judge's sentencing decision. Therefore, victims' decisions to report crimes enhance the likelihood that the goals of the criminal justice system will be met.

The decision to report crime has obvious importance not only to the system as a whole but also to the victim individually. A report to the police clearly indicates that a situation has arisen in the life of a victim that he or she believes deserves official action by police/law enforcement authorities. Such a decision is not necessarily unchangeable; a victim may report a crime and then later decide not to press charges or testify in court. Yet even when a victim vacillates after an initial report, the police may still make an arrest, and a prosecutor may still indict.

Until recently, victims' decisions concerning the reporting of crime were not regarded widely as a central step in the criminal justice process. Rather, it usually was supposed that the police were the initiators of the system. Currently, a victim's role in triggering the system has become more generally recognized; nevertheless, little is known about how victims make these critical decisions (Fattah, 1991). This chapter examines what motivates victims to report crimes to the police and investigates strategies that may increase victim reporting.

LITERATURE REVIEW

According to Ezzat Fattah (1991, pp. 43–45), a series of decisions, most often influenced by the victim, must take place before the criminal justice system becomes involved:

- The behavior in question must be noticed and defined as criminal. This discovery and definition may be made by a citizen (victim or witness) or by the police or prosecutor. Usually, the victim is first to make this decision.

- Someone must decide that a crime has occurred—that the behavior in question is properly within the scope of criminal law and the criminal justice system. Again, the victim usually determines this.

- A decision to enter the event into the criminal justice process ordinarily *requires that the event be reported to the police,* and in any case, it requires that the law enforcement establishment decides not only that a crime has occurred but also that the criminal justice process should be continued.

Although the police are generally considered the "gatekeepers" (see Chapter 2), perhaps victims should likewise be considered as gatekeepers, because it is they who ordinarily decide whether to report crimes to the police (Green, 1981; Skogan, 1984). This holds especially true with common crimes of theft and assault. Consumer frauds and victimless crimes, such as those involving gambling, narcotics, and prostitution, are less dependent on victim reporting, but in most cases, if the victim does not report a crime to the police, the criminal justice system will not become involved.

Until the development of the victimization surveys 30 years ago, there was no systematic data on either the extent or the characteristics of those crimes that remained unknown to the police. Victimization surveys, which have been administered by the Bureau of Justice Statistics since 1972, have revealed that a substantial proportion of victimizations reported to interviewers had not been reported to the police. For many kinds of crime, over half went unreported. For whatever reasons, vast numbers of victims decide not to become involved with the criminal justice process (Doerner & Lab, 1998).

A review of Bureau of Justice Statistics victimization data (Zawitz et al., 1993) reveals that violent crimes are more often reported than are household crimes such as burglary and theft. Other data suggest that crimes against businesses are more likely to be reported than are personal or household victimizations, that completed crimes are more likely to be reported than are those only attempted, and that crimes are more often reported when the victim has been injured.

Some evidence exists that age and sex may be factors in victim reporting. Victims of violent crimes who are age 12 to 19 are less likely to report them. Female victims of violent crimes are more likely to make police reports than are male victims (Zawitz et al., 1993).

Two methods have been used in an attempt to determine what influences victims' decisions whether to call the police. The first is to ask victims about it directly. The other is to analyze characteristics of the victim, offender, and event in terms of whether or not the police were called.

One advantage victimization surveys have over official crime statistics is the information they provide on the reasons why many victimization incidents, even serious ones, go unreported to the police (Fattah, 1991). In the 1960s, the National Opinion Research Center (NORC) conducted one of the earliest victimization surveys. The NORC survey asked people who reported a victimization to them but failed to notify the police for their reasons. The reasons they most commonly gave can be classified into four fairly distinct categories (Ennis, 1967, pp. 43–45):

- Thirty-four percent of those who did not report their victimization to the police failed to do so based on their belief that the incident was not a police matter. They did not want the offender to suffer or be harmed by the police, or they simply thought that the incident—usually an assault—was a private, and not a criminal, matter. Typical examples in this category were offenses committed by one family member against another or between friends.

- Two percent of the nonreporting victims feared reprisal, whether physically from the offender, his or her family or friends, or economically from either the cancellation of insurance policies or increases in premiums.

- Of those who failed to report, nine percent either did not believe it worth the time or trouble to get involved with the police, did not know whether they should call the police, or were too confused after the incident to do so.

- The greatest percentage, more than half, of the nonreporting victims failed to notify authorities because of their skepticism concerning the effectiveness of the police. Victims in this category believed that the police could do nothing about the incident, would not apprehend the offender(s), or would not want to be bothered. In short, these victims failed to report because of a negative attitude toward the police. This finding is of particular importance because it strongly suggests that an increase in police efficiency and an improvement in public attitudes toward the police are likely to result in increased reporting.

When victims do report criminal incidents, they generally cite crime prevention and punishment as reasons for calling the police. In the 1995 *Sourcebook of Criminal Justice Statistics,* more than a third of the respondents related that they reported personal crimes to catch or punish the offender. About one in five reported the victimization "because it was a crime" (Maguire & Pastore, 1996, p. 246). Victims of property crimes more often cited property recovery or insurance requirements as their reasons, but a fourth of these victims also felt obliged to report "because it was a crime" (Maguire & Pastore, 1996, p. 246).

In a study of 391 adult females who were interviewed about lifetime victimization experiences, Kilpatrick, Saunders, Veronen, Best, and Von (1987) found that over half of all crimes committed against these women were never reported to the police. Additionally, these researchers disclosed that reporting rates for sexual assaults were extremely low. Surprisingly, reporting rates of completed rapes were no higher than those of other sexual assaults. Kilpatrick and his colleagues (1987) also noted that less than half of all aggravated assaults were reported. In contrast, more than half of all robberies and burglaries were reported. Reporting rates were somewhat higher in more recent crimes, but nonreporting remained the norm for the most serious crimes. The findings in this study suggest that underreporting of major violent crimes is a significant problem, particularly for completed rapes and other forms of sexual assault. What Bachman (1998) did find was that rape victims are more likely to report this offense if they sustained physical injuries during the attack or if the offender used a weapon.

In other research to assess a victim's decision to notify the police, Greenberg, Ruback, and Westcott (1982) discovered that social influence was an important determinant in both the decision to call the police and the timeliness of such notification. If advised by others to call the police, the victim was more likely to report the crime. However, and somewhat ironically, the greater the number of people the victim consulted, the longer he or she delayed in reporting the crime. In initial studies of why police delay in responding to crimes, researchers found that the major source of the delays was not caused by police but by victims who often spoke with someone else before contacting the police (Van Kirk, 1978; Spelman & Brown, 1981). Additional findings from interviews with victims of rape, robbery, burglary, and theft indicated that a substantial proportion of them consulted with someone other than the police before reporting the crime (Ruback, Greenberg, & Westcott, 1984; Greenberg & Ruback, 1992).

Finally, Waller (1990) found that victim satisfaction leads to public support of the police and the potential of increased reporting. By apprehending an offender or recovering lost property and informing victims of these results, police assume a more favorable public image. Victims' satisfaction influences friends and family who may then be more likely to report subsequent victimizations. Similar results were found in Conaway and Lohr's (1994) longitudinal study utilizing National Crime Victimization Survey data. They likewise noted that crime victims were more likely to notify the police when a previous victimization had been reported to the police and the police conducted a routine follow-up investigation or when police activity in the previous victimization resulted in an arrest or the recovery of property. These results support theories that victims are more likely to cooperate with police if they have had positive experiences with them in previous victimizations.

VICTIMS' GOALS IN REPORTING VICTIMIZATIONS

It is easy to infer too much about how victims arrive at decisions to call or not to call the police by reviewing the data from the National Crime Victimization Survey (Zawitz et al., 1993). The data collected in these surveys are subject to the various limitations of the methodology used. According to Zawitz and his colleagues (1993), the main limitations may result from the following:

- Sampling (whether the interviewees adequately represent the whole nation)
- Accuracy of memory (how well the respondents remember the details of the incident)
- Willingness of respondents to participate
- Sensitivity of the subject matter
- Financial benefit of reporting
- Respondents' perceptions of the seriousness of the crime

The choices respondents are given on questionnaires are sometimes broad, ambiguous, and too widely open to personal interpretation. For example, the response "nothing could be done" might mean that after an assault, a victim believed that the physical harm done could not be corrected. Another victim may believe that his or her assailant could not be caught. Some victims may interpret this response as a belief that the criminal justice system remains powerless in preventing recurring attacks. The response "police would not want to be bothered" could mean that the victim thought the harm was very minor or that the police would simply not be interested in his or her case (Gottfredson & Gottfredson, 1988, p. 25).

Though not flawless, data from these surveys do suggest how victims exercise discretion and what factors may motivate them to report or not report a crime. Many victims tended to underestimate the crimes committed against them, assuming these incidents not worthy of official police/law enforcement action, whether because they themselves perceived them as unimportant or believed that the police would not be interested in pursuing them. This may represent the public's misconception that relatively minor crimes such as vandalism, a stolen bicycle, or verbal threats from a stranger fall outside the scope of the criminal justice system. The fact that completed crimes are more often reported than are those merely attempted is consistent with this interpretation.

Many of those who did report crimes to the police suggested utilitarian aims for reporting. Reasons such as "to catch the person," "to recover property," or "for personal protection" are examples. Others who reported an incident said they did so because they understood that a crime had been committed against them, a reporting incentive consistent with the idea that victims tend to seek validation of their feelings and perceptions. A general desire for crime prevention is also prominent as a reason given by many victims. Others indicate that their aim in reporting was retribution (Gottfredson & Gottfredson, 1988, p. 25).

Victims likewise reported a crime to the police for economic purposes. The relation between a victim's insurance status and his or her resolve to contact the police may explain the relatively high reporting rate of stolen automobiles. The decision to report a crime may be more influenced by a personal cost-benefit analysis than a civic obligation. Often, insurance companies will not refund victims until an official police report is made, and there is a strong association between reporting to the police and the victim's having insurance that covers loss of personal property. In one study, 85% of robbery victims with theft insurance reported the crime to the police, compared with only 51% of those without insurance (Gottfredson & Gottfredson, 1988).

One major characteristic of crime that affects a victim's report to the police is the nature of the offense suffered by him or her. The more serious the crime, whether measured in terms of physical injury or financial loss, the more likely it is that the crime will be reported. Victims are more likely to call the police for completed crimes than for attempts, for aggravated rather than simple assaults, for crimes involving weapons (particularly guns), for crimes causing injuries (particularly serious ones), and for crimes involving financial loss (particularly great loss). The severity of consequences associated with a crime offers a good predictor of whether the victim will call the police. Regardless of the type of crime, the more serious the harm, the more likely the crime will be reported to the police (Gottfredson & Gottfredson, 1988).

One might suppose that attitudes toward the police could substantially influence whether victims seek assistance from police officials after the commission of crimes, but studies have shown consistently that whether or not a victim reports an incident depends on the effects of the crime. Generally, when the harm done to victims is considerable, their attitudes toward the police do not play an influential role. The main determinant in reporting concerns the seriousness of the crime's consequences to the victim. Only when the crime involves little damage to the victim does his or her attitude toward the police influence reporting (Garofalo, 1977).

VICTIMS' DECISIONS TO NOTIFY THE POLICE

Greenberg and Ruback (1992) propose a theoretical model for reporting crimes to the police. According to Greenberg and Ruback (1992), the incentive for reporting victimizations to the police varies as a joint function of victims' level of distress and the belief that notification will improve their anguish. Research conducted by Greenberg and Ruback (1992) and others tends to support this reasoning. Findings from the British Crime Survey demonstrate that victims who notified the police were more emotionally affected by the crimes than those who did not report victimizations (Maguire & Corbett, 1987).

Other data reveal that the best predictor of reporting is related to how the victim perceives the seriousness of the crime (Fishman, 1979; Greenberg & Ruback, 1992; Schneider, Burcart, & Wilson, 1976; Skogan, 1984; Sparks, Genn, & Dodd, 1977; Waller & Okihiro, 1978; Webb & Marshall, 1989). Typically, victims report crimes more readily when the offenses have been completed rather than attempted (Bureau of Justice Statistics, 1990; Hindelang, 1976). Reporting increases when the value of property is estimated as high (Bureau of Justice Statistics, 1990; Schneider et al., 1976; Schwind, 1984; Waller & Okihiro, 1978) and when the crime has resulted in more significant injury to the victim (Block, 1974, 1989; Bureau of Justice Statistics, 1985). Similarly, in a study of rape victims in Atlanta, Greenberg and Ruback (1992) found that the victims who had suffered the greatest injury were more likely to report the incident. Higher rates of reporting crimes are also likely when victims perceive that the potential for greater harm existed, especially when the perpetrator possessed a weapon (Block, 1974; Bureau of Justice Statistics, 1985; Hindelang, 1976; Webb & Marshall, 1989), or during crimes in which threats were employed (Greenberg & Ruback, 1992).

Consistent with Greenberg and Ruback's (1992) proposed model, reporting crime is also related to feelings of being wronged or anger. Greenberg and Ruback's (1992) experimental studies identified a statistically significant difference between self-reported anger and willingness to call the police. Across the six experiments conducted by Greenberg and Ruback (1992), those who reported their victimizations were angrier, expressed a greater desire to see the offender punished, and felt greater moral obligation to report than nonreporters. Additional evidence linking anger and reporting was also obtained in Greenberg and Ruback's (1992) analysis of rape victims in Atlanta. However, Waller and Okihiro's (1978) study of burglary victims found no relationship between anger and reporting. Their failure to find a relationship may have been due in part to the long time interval between the burglary and the interview, up to three years in some cases.

Evidence also supports the facilitating effect of fear and defenselessness on reporting. Dukes and Mattley (1977) found that rape victims who reported their attacks, as opposed to nonreporters, were more worried about reprisal and more strongly motivated to prevent a recurrence of the attack. Examination of the reasons offered for reporting crimes lends additional weight to feelings of vulnerability and apprehension as sources motivating the decision to notify police. Indeed, a major reason offered by victims who did report was their desire to prevent the crime from ever reoccurring (Bureau of Justice Statistics, 1990; Greenberg & Ruback, 1992; Waller & Okihiro, 1978).

Greenberg and Ruback (1992) suggest that it is not only the magnitude of the victim's distress that motivates reporting the crime, but also the belief that reporting is likely to reduce distress. Support for this proposition comes from a study by Dukes and Mattley (1977), who found that rape victims were most likely to report the crime when they were fearful and anticipated a positive response from the police. Similarly, Schneider et al. (1976) found that beliefs in police effectiveness were related to the reporting of serious victimizations but not minor property crimes. Additional support for the importance of expecting assistance as a motive for calling the police can be found in the reasons offered by victims for not notifying the police. Results from the National Crimes Panel indicate that the most frequently given reasons for not summoning the police are victims' beliefs that nothing could be done and that the police would not want to be bothered with investigating the complaint (Law Enforcement Assistance Administration, 1982; Mawby & Walklate, 1994). Similar studies conducted by Waller and Okihiro (1978) and Greenberg and Ruback (1992) support such findings.

What do victims hope to gain by notifying the police? According to Greenberg and Ruback's (1992) model, victims are motivated to reduce the feeling of being wronged and of being vulnerable to subsequent victimization. Victims may be mobilized to report the crime because they believe that the police will be able to apprehend the offender and, in the case of property crimes, recover the stolen property. Justice would also be achieved if the offender were convicted and punished and/or forced to make restitution to the victim. The latter goal has become increasingly more feasible in recent years as a result of the judiciary's more favorable attitudes toward restitution (Chesney, Hudson, & McLagen, 1978).

Even if victims believe it is unlikely that the perpetrator will be brought to justice, they may still anticipate positive consequences from reporting a crime. For example, the crime must be reported to the police if the victim is to receive financial reimbursement from

third parties such as insurance companies and state-sponsored victim compensation programs (Cain & Kravitz, 1978).

Notifying the police can also reduce victims' fears of future victimizations. If the suspect is arrested and subsequently incarcerated, then he or she no longer poses a threat to the
victim. Moreover, victims may believe that such punishment will deter potential offenders.
Even when a victim sees little likelihood that police will apprehend the offender, he or she
may perceive that informing the police of a crime will lead to more intensive police pa-
trolling, which, in turn, serves to reduce the victim's fear of future susceptibility to criminal attack.

Though victims may anticipate numerous positive consequences from calling the police, the fact remains that little more than a third of victimizations are reported. This statistic suggests that many victims may be pessimistic about the outcomes of reporting a crime
or that they expect additional emotional, material, or psychological costs from their involvement with the police and other personnel involved in the criminal justice system (Cretney & Davis, 1995). Research discloses a number of potentially negative consequences of
victim involvement with the police and the criminal justice system, including the expense
of transportation and parking, the loss of time from work or school, and the additional cost
of child care (Knudten, Meade, Knudten, & Doerner, 1976). Questioning by the police and
opposing attorneys can also be stressful to victims. Victims of rape, in particular, often feel
embarrassed and frustrated as a result of their encounters with the police and courts and
sometimes conclude that they are the ones on trial (Amir, 1971; Binder, 1981). In addition,
fear of retaliation by the perpetrator, although infrequently cited in victimization surveys,
may nevertheless be an important cost for certain classes of victims. In one survey, 75% of
racial minority victims of completed rapes indicated fear of reprisal as their reason for not
notifying the police (Hindelang & Davis, 1977).

These studies suggest that the anticipated negative consequences of calling the police
outweigh the anticipated benefits. Indeed, one reason frequently cited for not reporting is
the belief that a crime was not important enough to request outside assistance and that getting the police involved was a nuisance (Bureau of Justice Statistics, 1990; Mawby & Walklate, 1994). The following two subsections further explore what may motivate victims in
their decision to report or not report crimes to the police.

When Victims Report Crimes to the Police

According to the 1984 British Crime Survey, 33% of all crime victims indicated a sense of
obligation as the underlying reason for deciding to go to the police (Hough & Mayhew,
1985); they believed that reporting the offense was *the right thing to do* (Shapland, Willmore, & Duff, 1985). Victims expressing this sense of social obligation may be motivated
to protect others from being victimized. Likewise, they may report a crime as a matter of
routine or expect that their reports will increase police patrols. The inconvenience of reporting, even when considerable, seems offset by one's perception of civic responsibility to
his or her fellow citizens (Gottfredson & Gottfredson, 1988).

Another motive for reporting crimes, particularly in assault cases in which the victim
knew the assailant, is the victim's desire to see his or her assailant confronted in a public
forum. For many victims, it is this, rather than anticipation of the court's sentence, that motivates them to report a crime to the police. Public recognition that a crime has occurred

seemingly vindicates the victim simply by the criminal justice system's acknowledgment of the perpetrator's act. This may be especially true in cases of domestic violence (Cretney & Davis, 1995).

A third reason why assault victims choose to go to the police is their fear of repeat attacks. They wish to ensure some control over their assailants' immediate behavior and protect themselves from further victimization. Once again, this seems especially true in assault cases in which the victim and assailant are acquaintances. Unfortunately, victims who go to the police for protection do not always receive the help that they think they deserve. Though vulnerable, they do not necessarily attract police sympathy—perhaps the police view them as *willing victims,* either because they participate in lifestyles viewed as socially deviant (prostitutes, drug dealers, addicts, gang members) or are held to be of little value in the currency of the courts. In some cases, such as domestic violence, victims may repeatedly call upon the police for assistance but then refuse to cooperate in having their assailants arrested. Clearly, they desire the immediate victimization to stop, but they may not take the necessary actions to bring about a more permanent resolution to their situation. As a result, police officers may not respond in a timely manner to those situations and circumstances in which they feel a victim has shown a prior reluctance to press charges because of his or her relationship to the assailant (Bureau of Justice Statistics, 1985, 1990, 1994; Cretney & Davis, 1995).

Another possible motive for reporting crime is the victim's wish to secure financial compensation. The victim of a property crime may be required by his or her insurance carrier to make a formal police report before filing a claim. In some cases, police officers, skeptical of this motive for reporting crime, suspect that a desire to secure a financial award may prompt "unworthy" reports. On the other hand, financial retribution is rarely a factor in reporting person offenses (McDonald, 1976). In their study of assault victims, Cretney and Davis (1995) support the finding that financial compensation was not a factor for reporting victimization.

When Victims Do Not Report

Reasons most often given for not reporting a crime to the police are that the crime was a private or personal matter, that the loss was recovered, that the crime was merely attempted and not successfully completed, or that the consequences of the offense seemed insignificant. Numerous crimes go unreported because victims perceive that the police do not want to be bothered with crimes that are not considered serious. In typical surveys, interviewers have determined that nearly one fifth of the rapes, two fifths of the robberies, three fifths of the larcenies, and one half of the household burglaries reported to them were not reported to the police because the victims either believed that nothing could be done or that the offense was not important enough. Additionally, victims of rapes and assaults have indicated in surveys that they did not report the event to the police because they believed that the victimization was a private matter. For some types of crimes (burglary, larceny, robbery), fear of reprisal is typically not mentioned as a reason for not reporting the crime to police. For rape victims, however, victim surveys reveal they may not report their victimization to police because they do fear future retaliations (Zawitz et al., 1993).

According to Cretney and Davis (1995), a victim's perception of the detectability of an offense strongly influences his or her reporting behavior. Some victims choose not to report an incident because they cannot identify the perpetrators and therefore see no prospect of the crime being solved (Hough & Mayhew, 1985). Assault victims appear to have rather low expectations of the police; they understand only too well that "real" police do not share the same success rate in arresting suspects as the fictional detectives found in novels or on television shows. In circumstances where they themselves are unable to assist in the identification of their perpetrators, victims may lack any confidence that their assailants will be apprehended (Cretney & Davis, 1995).

Another reason cited for nonreporting is a victim's refusal to acknowledge a "victim" identity; some people regard going to the police as synonymous with weakness. Asking for help may conflict with an individual's sense of personal strength or endurance. It may likewise interfere with an individual's need to deny his or her victimization. As an example, a victim of domestic violence may wish to keep his or her victimization private; a wife's embarrassment or a husband's shame may keep either from seeking police intervention. Some studies support the position that battered women may not consider themselves as victims (Hough & Mayhew, 1985).

In addition to victims of domestic violence, young men may be particularly likely not to report victimization as a result of their trying to maintain an image of rugged masculinity. After being assaulted in nightclubs by strangers, they may continue to frequent such places, believing that they can look after themselves. Perhaps feeling that there is no realistic chance of detection affects their decision not to go to the police; however, by exhibiting a degree of self-confidence and self-reliance they appear to either reject the "victim" label or dismiss the incident as too minor to require police intervention (Cretney & Davis, 1995).

In the 1988 British Crime Survey, fear of reprisals was given as a reason for nonreporting in 2% of common assaults, 3% of other assaults, and 5% of woundings (Mayhew, Elliott, & Dowds, 1989). The U.S. National Crime Survey Panel found that 4% of victims gave fear of reprisal as their reason for not reporting an assault, although the proportion doubled in more serious cases (Hindelang, Dunn, Sutton, & Aumick, 1975). Similarly, the U.S. National Crime Victimization Survey disclosed that 4% of assault victims and 4.5% of those who were victims of more serious crimes gave fear of reprisal as a reason for not reporting their victimizations to the police (Bureau of Justice Statistics, 1994).

According to Cretney and Davis (1995), survey data significantly underrepresent the degree to which assault victims are deterred from going to the police because they fear reprisals. Acknowledged by 14% of the victims Cretney and Davis (1995) interviewed, fear of reprisal was especially prominent when the victim and assailant had an ongoing relationship—whether as partners, family members, or as members in some type of organized crime affiliate. In situations where there exists no ongoing relationship marked by a power imbalance between victim and assailant, the victim's course of action seems to be straightforward—report the matter to the police and assist in any way possible with the prosecution process. But where there is an ongoing relationship, as in domestic violence, there may be no effective way to negate the victim's fear of reprisal. Even when police are enthusiastic and there is the potential of a successful prosecution, the victim may be reluctant to cooperate because of intimidation.

Kidd and Chayet (1984) indicate that victims fail to report crime because they are concerned about their own safety, because they are threatened by future victimization, or because they believe that the police can do little to resolve the crime. Police may not be called as well in situations in which individuals feel vulnerable because of self-recrimination and loss of personal control (Wortman, 1983). This is especially true in sexual assault cases. In those instances, the female victim may feel that she has, through her own actions, precipitated the crime. This reason for not reporting may be particularly detrimental to victims of sexual assault as research indicates that victim recovery requires support from others, including police/law enforcement personnel (Sales, Baum, & Shore, 1984).

Victims who fail to report crimes, particularly victims of sexual assaults, have been characterized by feelings of despair. Their fear and powerlessness dictate that they suffer their victimizations quietly. The emotional costs of being a "silent" crime victim can be especially severe, and current research has alerted police/law enforcement personnel to the multiple and complex psychological effects that all victims experience. In light of this information, the criminal justice system has been called to address the needs of victims with greater sensitivity. Insights into why victims of sexual assault may fail to report these crimes, in addition to the effects of such crimes, have influenced discussions concerning the reporting of other types of crime as well.

According to Skogan (1984), the National Crime Survey in 1981 showed that only 56% of robberies and 51% of burglaries were reported to the police. Fattah (1991) noted that the Canadian Urban Victimization Survey also found high levels of nonreporting for all crime, including violent offenses. The Canadian survey indicated, as have other studies, that the most common reasons given for failure to report an offense are that the incident was "too minor," that the police could do nothing about the incident, and that it was too inconvenient for the victim. Of note in these findings is that victims of sexual assault frequently had a negative attitude toward the police and therefore were reluctant to report the incident (Fattah, 1991).

Individuals may sometimes feel that by ignoring offenses or by "handling" the offenses themselves, they have no need to pursue police assistance. Skogan (1984) asserts that under certain circumstances, cases might not need authoritative or official intervention. As an example, one may understand why a parent is reluctant to report a child's criminal activity (theft or destruction of household property, minor assault) to police when the consequences of the child's offense have only affected his or her family. Skogan (1984) has conducted small surveys in which respondents reported having taken care of these sorts of problems themselves. This "self-help" approach to dealing with a criminal incident is often given as a reason for not involving the police. Referring to it as "rough justice," Skogan (1984, p. 129) has called for more research concerning this form of informal dispute resolution, which may involve direct confrontation with the offender and result in immediate restitution or retribution. It may include mediation through a third party who plays a role in reducing conflict and arbitrating a settlement. In the minds of some victims, informal response to crime may serve as a reasonable alternative, bringing about a resolution and reducing the necessity to pursue any further actions through formal channels.

The standard approach in victimization surveys is to establish whether or not a crime has occurred and, if so, what happened. The focus then turns to whether or not the crime was reported to the police. Recent studies measuring the reporting practices of victims have altered perceptions about victimology in regard to what has been called the lack of "formal" response to criminality. As Block and Block (1984) noted, decisions made by the criminal justice system have, in the past, been more intensively studied than have decisions made by victims. This has created an impression of an unresponsive system, a system that ignores a substantial proportion of crimes, a system that allows victims to feel alone in their attempts to address whatever injustices have been directed toward them. Therefore, it is up to the system to help enhance victims' reporting behaviors.

STRATEGIES TO INCREASE REPORTING

The significance of the police role in responding to crime victims cannot be overemphasized. Perhaps the most valuable strategy police can use to increase reports of victimization is to ensure that those who *do* report crimes are treated justly and with dignity. Police officers interact more often with crime victims than other professionals in the criminal justice system. The manner in which dispatchers, the first officers who arrive at the scene, and the detectives who will investigate the case treat a victim all shape expectations of future treatment throughout the justice process. It is therefore critical that every police/law enforcement professional who comes into contact with crime victims, either in person or over the telephone, knows how to respond in a sensitive and effective manner.

First responders, such as sheriff and police departments, must ensure that victims receive adequate care, essential information, and emergency assistance in the immediate aftermath of a crime. State and federal laws mandate victims' rights that police/law enforcement officials must fulfill. These laws generally include the following: information about crime victim compensation; information about referrals to various types of victim services; protection from intimidation and harm, including aggressive enforcement of antistalking and restraining orders; information about the status of the investigation; and notification when the accused is released from custody (Greenberg & Ruback, 1992).

The Department of Justice has long recognized the significant role policing/law enforcement plays in providing information and assistance to victims of crime. Over 20 years ago, policing/law enforcement was one of the first allied professional groups targeted for training and technical assistance by the Law Enforcement Assistance Administration (LEAA) within the Justice Department. LEAA helped fund innovative programs including assistance to elderly victims and mobile crisis units with police officers and mental health professionals to respond to crime scenes. During that time, the first police-based victim assistance programs were established in Ft. Lauderdale, Florida, and Indianapolis, Indiana (Doerner & Lab, 1998).

The President's Task Force on Victims of Crime (1982) provided a major impetus for the victims' movement in the United States. After traveling to six cities and hearing testimony from over 1,000 people, the task force presented its report and 68 recommendations for improving the treatment of crime victims. Implementation of the recommendations was fostered by the passage of the Comprehensive Crime Control Act in 1984. One of its provisions, the Victims of Crime Act (VOCA), made federal funds available to states with compensation programs. The act also authorized the disbursement of funds through the Office of Victim Assistance, an agency within the Department of Justice, to aid victim agencies that had previously relied heavily on volunteers (Office of Justice Programs, 1986). The Office of Victims of Crime supports 113 victim assistance programs in police departments across the country through VOCA funding. Many departments are using these funds to serve victims in innovative ways. For example, the Detroit, Michigan, Police Department uses VOCA money to fund a rape council, which provides hospital accompaniment, counseling, criminal justice advocacy, and other vital services to victims of sexual assault (Office of the Attorney General, 1995).

The President's Task Force on Victims of Crime (1982) identified four important areas for improving policing/law enforcement's treatment of crime victims. Implementation of these may increase the reporting of victimizations:

- Training programs to increase sensitivity and awareness about victims' issues
- Prompt property-return procedures
- Periodic information to victims regarding the status of their case and the closing of the investigation
- Priority on investigating threats or intimidation

To varying degrees, federal, state, and local law enforcement agencies have made progress in these areas. However, there exists no comprehensive data on the percentage of law enforcement agencies that provide basic victim assistance services. According to the International Association of Chiefs of Police, the majority of large city police departments have established victim assistance programs; however, the majority of police/law enforcement agencies serving smaller jurisdictions and rural areas have not (Weisheit, Falcone, & Wells, 1994). In a Bureau of Justice Statistics (1995) survey of police agencies with more than 100 sworn personnel, 37% of responding agencies reported that they operated special victim assistance units.

The range of services provided by police agencies across the nation varies significantly. A growing number of agencies employ full-time advocates to provide comprehensive assistance to victims, and many agencies have adopted policy statements on the basic level of assistance victims should receive. Numerous police departments aid victims by giving them brochures on community-based victim assistance programs, the legal system, and the needs of specific victim groups, such as those affected by domestic violence and sexual assault (Bureau of Justice Statistics, 1995).

According to Beatty, Smith-Howley, and Kilpatrick (1997), the Austin, Texas, Police Department developed a comprehensive counseling and victim assistance program with three specialized units: day service, child abuse, and family violence. The department uses a special on-scene crisis unit; equipped with unmarked police cruisers and radios, counselors assist victims by making alternative living arrangements when a spouse and children need to leave abusive situations. Funding is available to pay for up to a two-night stay in a hotel. The department also operates a substation that is staffed by neighborhood officers, a counselor from the department's victim services division, and a city services coordinator. These substations are mobile vehicles, usually vans, located in high-crime neighborhoods.

In Kentucky, the Lexington Police Department created the first automated system in the nation to notify victims of any pretrial/preconviction release of crime perpetrators from custody or when there is a change in their status. The system, which is now used throughout the state, calls victims until a successful notification is made, giving them time to take safety precautions if necessary. The system gives the public, including victims, access to critical inmate information 24 hours a day, seven days a week. Over 300 state and local jurisdictions are in the process of implementing similar notification systems (Beatty et al., 1997).

In Waltham, Massachusetts, and Skokie, Illinois, police departments have installed panic devices in the homes of at-risk victims in response to the growing number of stalking and domestic violence incidents reported in those states. When activated, these devices signal the police department through wireless transmitters. Many police/law enforcement agencies have taken stalking and domestic violence prevention even further by arranging the free use of cellular phones by victims, giving them greater mobility (Beatty et al., 1997).

Federal law enforcement agencies have taken major steps to enforce victims' rights and to increase reporting. Under federal law, agencies are responsible for identifying victims and witnesses of federal crimes, informing victims of their rights immediately after an incident, and referring them to emergency medical and victim services. These and other important responsibilities are set forth in the *Attorney General Guidelines for Victim and Witness Assistance,* which stresses that, whenever possible, police/law enforcement officials must assist victims in making contact with appropriate medical or victim-related services (Office of the Attorney General, 1995).

More than 70 federal agencies have a police/law enforcement function, including the Federal Bureau of Investigation (FBI), military agencies (i.e., Naval Criminal Investigative Service, Air Force Office of Special Investigations), the Bureau of Indian Affairs, U. S. Customs Service, U.S. Postal Inspection, and the U.S. Park Service. These agencies are reaching out to crime victims by developing informational brochures in various languages, conducting national and regional training programs for their investigators, and designating victim-witness coordinators in their offices (Office of the Attorney General, 1995).

Policing/law enforcement has taken an active and pivotal role in the development of multidisciplinary team approaches to crime, first in response to child abuse in the 1980s, and then in response to sexual assault and domestic violence in the 1990s. Many police departments have established special multidisciplinary programs, which improve their response to victims by utilizing the expertise of numerous professionals in one setting (see Chapter 5). In Tennessee, the Nashville Police Department created a Victim Intervention Program in 1975. The unit, now staffed by mental health counselors, provides free crisis intervention and ongoing counseling for any victim who has been affected emotionally. In 1994, the department expanded its assistance to victims by creating a separate Domestic Violence Intervention Division that coordinates its response with the prosecutor's office. The largest program of its kind in the nation, the division is staffed by more than 32 specially trained professionals who handle thousands of cases each year. The Nashville Police Department credits this unique intervention with helping to reduce domestic violence homicides by over 40% during a two-year period (Beatty et al., 1997).

The New Haven, Connecticut, Department of Police Services and the Child Study Center at Yale University School of Medicine have developed a unique collaborative program to address the psychological impact of chronic exposure to violence on children and families. The initiative, called the Child Development–Community Policing Program, brings together police officers and mental health professionals who provide each other with training, consultation, and support, as well as direct interdisciplinary intervention to children who are victims, witnesses, or perpetrators of violent crime (Beatty et al., 1997).

In Florida, the Largo Police Department collaborates with the state attorney's office and with the local domestic violence shelter to enhance services and to improve the prosecution rate of domestic violence cases. Staff in the department's domestic violence unit receive intensive training on responding to domestic violence incidents more effectively. The unit uses technological advances to help gather evidence, including lapel microphones to record audio arrival at the scene and camcorders to videotape victim and witness statements on the scene. One of the unit's most innovative services is faxing copies of police reports to the local shelter, which then calls victims to offer assistance (Beatty et al., 1997).

In many communities, police officers and health care professionals are using a new approach to handling sexual assault and child abuse cases in which victims are treated in specialized settings more hospitable than emergency rooms. In these secure and comfortable environments, trained nurse examiners conduct evidentiary medical exams. For example, one such program in Tulsa, Oklahoma, which has been recognized by the Ford Foundation and the Kennedy School of Government as an important public sector innovation, was initiated through the efforts of local police and medical professionals. Sexual assault victims are treated in a quiet, comfortably appointed section of the hospital. This setting provides a welcoming and supportive environment for victims. After conducting more than 500 rape examinations in this special setting, Tulsa police officials reported substantial improvement in the quality of forensic evidence and a higher rate of convictions due to a greater willingness of victims to undergo examinations. Tulsa officials also reported that all cases in which nurses provided testimony resulted in convictions. Additionally, because of this innovative program, Tulsa police noted an increase in reporting sexual assaults (Beatty et al., 1997).

The concept of community policing may also increase the reporting behavior of victims. Community policing is based on a philosophy of problem solving at the local level. Both police officers and private citizens work together in creative ways to remedy contemporary community problems related to crime, fear of crime, social and physical disorder, and neighborhood decay (Carter, 1995). Community policing supports the belief that police departments must develop new relationships with the citizens they serve by allowing community members to have a voice in setting local law enforcement priorities and involving them in efforts to improve the quality of life in their neighborhoods (Carter, 1995). Rather than spending a majority of their time responding to random calls, police approach local criminal activity proactively with the help of community members (Carter, 1995).

The ultimate goal of community policing is to reduce crime by using community-police partnerships to develop crime prevention strategies that work. As police officers develop trust with residents in neighborhoods, community policing may encourage victims who traditionally do not report crimes to participate in the system and seek assistance for their financial, physical, and emotional injuries.

The Violent Crime Control and Law Enforcement Act of 1994 authorized funds to promote community policing, in addition to adding another 100,000 police officers to the ranks over a six-year period. The Office of Community Oriented Policing Services (COPS) within the U.S. Department of Justice is responsible for carrying out this mission, and as of December 31, 1999, the agency has awarded grants for the hiring or redeployment of more than 101,000 police officers and sheriff deputies. The COPS Youth Firearms Violence Program supports innovative community policing approaches to fight firearm violence among young people, and the COPS Community Policing to Combat Domestic Violence Program provides grants to local communities to battle domestic violence.

According to research sponsored by the National Institute of Justice, community policing has been adopted in many jurisdictions across the nation. In a 1993 survey of 2,314 municipal and county police and sheriff departments, nearly 20% of the respondents had implemented a community policing approach, and 28% were in the process of doing so. The same study showed the nationwide benefits community policing produces. Among agencies that had implemented community policing for at least a year, 99% reported improved cooperation between citizens and police, including *increased reporting of victimizations.* Eighty percent reported a reduced fear of crime among the residents in their communities, and 62% reported fewer crimes against persons (Office of Community Oriented Policing Services, 1997a).

Other innovative uses of collaborative police-community programs may assist in encouraging victim reporting. In San Diego, California, the police department, in partnership with the YWCA of San Diego County, has created a Community Domestic Violence Resource Network. With support from a $200,000 COPS grant, a toll-free telephone clearinghouse provides victims access to information on all domestic violence service providers in the county. In addition, this network has proven to be a particularly valuable resource for community police officers in their response to domestic violence calls. Using a computerized database, specially trained information specialists can now offer law enforcement officers immediate information concerning which shelters have space available, which accept children, and other relevant information (Office of Community Oriented Policing Services, 1997b).

In Provo, Utah, the sheriff's office reports that they have dramatically increased the human resources available to solve crime problems by involving victims in victim-assisted investigations. When appropriate, officers enlist the participation of crime victims by explaining the type of information needed to make an arrest and guiding them on the role they can play in the department's investigation (Office of Community Oriented Policing Services, 1997b).

In Lowell, Massachusetts, police report that for the first time in 25 years, a year passed without a single murder. This extraordinary event, police believe, is a product of intensive community policing. With support from a COPS grant, the city hired 65 new officers, created bicycle patrols, and implemented a gang unit. The department also created neighborhood police storefront substations that serve as a base of operations for community police officers assigned to neighborhood foot patrols. Overall, crime has declined in all but one category; there has been an increase in reported assaults. However, this statistic, the department believes, is the direct result of increased victim reporting. Community policing efforts have helped women feel more comfortable and have positively influenced their decisions to report domestic violence offenses (Office of Community Oriented Policing Services, 1997c).

A number of police departments have implemented innovative programs offering multilingual, multicultural support to victims and survivors of gang violence. Much like domestic violence victims, victims and witnesses of gang violence generally live with or among the perpetrators of the crimes and, fearing retaliation, are especially reluctant to participate in the system. These programs are usually collaborative efforts, with victim advocacy organizations and police departments providing comprehensive services for victims of gang violence. Services offered by these programs include emergency crisis response, sensitive death notification, accompaniment of survivors to emergency rooms, intervention with employers and the media, assistance in obtaining victim compensation, referrals for counseling services, and training for emergency medical and hospital personnel in responding to gang victims more effectively and sensitively (Office of Community Oriented Policing Services, 1997b).

Like their civilian counterparts, the military has also instituted programs and guidelines to ensure that victims are treated with sensitivity and provided essential information and emergency assistance in the aftermath of victimization. The Department of Defense

(1994) mandates that victims should be informed of available crime victim compensation, available command and/or community-based victim treatment programs, the stages in the military criminal justice process, and how the victim can obtain information about the status of his or her case. Through the military's community policing and crime prevention programs, the Department of Defense conducts regular victimization awareness training with military personnel and their dependents in an effort to sensitize this population to the crime problem and to *increase crime reporting.* Much of this training is sponsored by Officer Wives Clubs and the Ombudsmen, an enlisted wives support organization (Department of Defense, 1994).

Studies have shown that victims are often reluctant, for a variety of reasons, to report crimes. Police and law enforcement agencies, having recognized that crime victims are the first link to solving crimes and reducing criminal activity, are working harder than ever to increase victims' reporting behaviors. As community involvement becomes a dominant force in policing, police/law enforcement agencies should enhance the ways they incorporate victims of crime into their community partnerships. Because victims are often dedicated to solving cases in which they or their families have been injured, they may also serve as valuable catalysts by organizing additional community-police initiatives, joining existing problem-solving efforts, and/or providing key information to investigators.

CONCLUSION

Typically, police do not discover crimes. Without victim reports, the criminal justice system falters. Research validates that victims can no longer be ignored throughout the system if the system wishes to succeed in its investigative and crime prevention mission. Therefore, encouraging victims to report crimes has moved into the forefront of policing strategies.

Understanding why victims may fail to report crimes is the first step in advancing their role in the criminal justice system. Studies have shown that certain types of crimes are more likely to be reported than others. Violent crimes tend to be reported more often than property crimes. The more serious the crime, in terms of the victim's perception of injury or loss, the more likely he or she is to report it to the police.

Reasons most frequently cited by victims for *not* reporting crimes to police/law enforcement officials include a belief that the event was a personal matter. This is particularly true for unreported rapes and other assaults. Attempted crimes often go unreported as well as those crimes with which victims suspect the police would not want to be bothered. Failure to report may result from a victim's perception that the crime was not important or that nothing could be done about it.

One way to increase victims' reports is to enhance the relationship between local police agencies and the communities they serve. Equally important is the assurance of just and sensitive treatment for those victims who do decide to report crimes. Recently, innovative police-community programs have worked to enhance their services to victims by including mental health, community resource, and medical professionals in a team approach to aiding victims. These initiatives have resulted in enhanced police-community relations and increased reporting behaviors. Clearly, this suggests that there is a close connection between actual victims' and potential victims' perceptions of the police and their decisions to report or not report crimes.

Because victims initiate the criminal justice process, they should be regarded as the most influential decision makers in the system. Minimizing their role in the criminal justice process will not encourage future reporting. Generally, victims who do report crimes express that the prevention of future criminal activity served as an incentive to do so, an indication that victims do wish to participate more actively in the criminal justice process. Community-based policing makes good use of citizens' desire to feel safe in their neighborhoods. By involving potential victims in their own protection and ensuring that should they be victimized, the police will respond in a genuinely caring manner, police/law enforcement agencies build trust with the people they serve and establish an environment that encourages reporting.

Many police/law enforcement agencies and officials are responding to crime and enhancing victim reporting behaviors with an unprecedented level of comprehensiveness. They have made a serious commitment to our society's most vulnerable victims and potential victims through special units devoted to investigating domestic violence, sexual assault, and child abuse; by forging unique partnerships with elderly citizens; and by incorporating victim advocacy in community policing efforts. Despite this record of progress, however, more work needs to be done. The police/law enforcement response to victims of crime remains inadequate in some communities around the country. Crime victims still report that law enforcement officers have failed to be sensitive or provide them with critical information. Some multilingual and multicultural communities may not have adequate resources, training, or support programs in place to serve victims, thereby neglecting the needs of those who have already been victimized and minimizing the likelihood of future reporting.

Until recently, the role of victims in the criminal justice system has typically been neglected by police and ignored by researchers. Additional studies measuring victim reporting behaviors, innovative programs, and continued efforts to enhance police-community partnerships will undoubtedly aid law enforcement agencies in strengthening their links to the most vital players in the criminal justice system—victims.

REFERENCES

AMIR, M. (1971). *Patterns in forcible rape.* Chicago: University of Chicago Press.

BACHMAN, R. (1998). The factors related to rape reporting behavior and arrest: New evidence from the National Crime Victimization Survey. *Criminal Justice and Behavior, 25*(1), 8–29.

BEATTY, D., SMITH-HOWLEY, S., & KILPATRICK, D. G. (1997). *Statutory and constitutional protection of victims' rights: Implementation and impact on crime victims.* Arlington, VA: National Victim Center.

BINDER, R.L. (1981). Why women don't report sexual assault. *Journal of Clinical Psychiatry, 42*(4), 427–438.

BLOCK, C. R., & BLOCK, R. L. (1984). Crime definition, crime measurement, and victim surveys. *Journal of Social Issues, 40*(1), 137–160.

BLOCK, R.L. (1974). Why notify the police: The victim's decision to notify the police of an assault. *Criminology, 11*(4), 555–569.

BLOCK, R.L. (1989). Victim-offender dynamics in stranger to stranger violence: Robbery and rape. In E.A. Fattah (Ed.), *The plight of crime victims in modern society* (pp. 231–251). London: Macmillan.

Bureau of Justice Statistics. (1985). *Reporting crimes to the police.* Washington, DC: U.S. Department of Justice.

Bureau of Justice Statistics. (1990). *Criminal victimization in the United States: 1988.* Washington, DC: U.S. Department of Justice.

Bureau of Justice Statistics. (1994). *Criminal victimization in the United States: 1992.* Washington, DC: U.S. Department of Justice.

Bureau of Justice Statistics. (1995). *Law enforcement management and administrative statistics, 1993: Data for individual state and agencies with 100 or more officers.* Washington, DC: U.S. Department of Justice.

CAIN, A.A., & KRAVITZ, M. (1978). *Victim/witness assistance: A selected bibliography.* Washington, DC: U.S. Government Printing Office.

CARTER, D.L. (1995). *Community policing and D.A.R.E.: A practitioner's perspective.* Washington, DC: U.S. Department of Justice.

CHESNEY, S., HUDSON, J., & MCLAGEN, J. (1978). New look at restitution: Recent legislation, programs, and research. *Judicature, 61*(8), 348–357.

CONAWAY, M.R., & LOHR, S.L. (1994). A longitudinal analysis of factors associated with reporting violent crimes to the police. *Journal of Quantitative Criminology, 10*(1), 23–39.

CRETNEY, A., & DAVIS, G. (1995). *Punishing violence.* New York: Routledge.

Department of Defense. (1994). *DOD directive 1030.2: Victim and witness assistance.* Washington, DC: Author.

DOERNER, W.G., & LAB, S.P. (1998). *Victimology* (2nd ed.). Cincinnati, OH: Anderson.

DUKES, R.L., & MATTLEY, C.L. (1977). Predicting rape victim reportage. *Sociology and social research, 62*(1), 63–84.

ENNIS, P.H. (1967). *Criminal victimization in the United States: A report of a national survey.* Washington, DC: U.S. Government Printing Office.

FATTAH, E.A. (1991). *Understanding criminal victimization.* Englewood Cliffs, NJ: Prentice-Hall.

FISHMAN, G. (1979). Patterns of victimization and notification. *British Journal of Criminology, 19*(2), 146–157.

GAROFALO, J. (1977). *The police and public opinion.* Washington, DC: National Criminal Justice Information and Statistics Service.

GOTTFREDSON, M.R., & GOTTFREDSON, D.M. (1988). *Decision making in criminal justice: Toward the rational exercise of discretion.* New York: Plenum Press.

GREEN, G.S. (1981). *Citizen reporting of crime to the police: An analysis of common theft and assault.* Ph.D. dissertation, University of Pennsylvania, Philadelphia.

GREENBERG, M.S., & RUBACK, R.B. (1992). *After the crime: Victim decision making.* New York: Plenum Press.

GREENBERG, M.S., RUBACK, R.B., & WESTCOTT, D.R. (1982). Decision making by crime victims: A multimethod approach. *Law & Society Review, 17*(1), 47–84.

HINDELANG, M.J. (1976). *Criminal victimization in eight American cities: A descriptive analysis of common theft and assault.* Cambridge, MA: Ballinger.

HINDELANG, M.J., & DAVIS, B.J. (1977). Forcible rape in the United States: A statistical profile. In D. Chappell, R. Geis, & G. Geis (Eds.), *Forcible rape: The crime, the victim, and the offender* (pp. 87–114). New York: Columbia University Press.

HINDELANG, M.J., DUNN, C.S., SUTTON, L.P., & AUMICK, A.L. (1975). *Sourcebook of criminal justice statistics, 1974.* Washington, DC: U.S. Government Printing Office.

HINDELANG, M.J., & GOTTFREDSON, M.R. (1976). The victim's decision not to invoke the criminal process. In W.F. McDonald (Ed.), *Criminal justice and the victim* (pp. 57–78). Beverly Hills, CA: Sage.

HOUGH, M., & MAYHEW, P. (1985). *Taking account of crime: Key findings from the 1984 British crime survey.* London: Her Majesty's Stationary Office.

KIDD, R.F., & CHAYET, E.F. (1984). Why do victims fail to report? The psychology of crime victimization. *Journal of Social Issues, 40*(1), 39–50.

KILPATRICK, D.G., SAUNDERS, B.E., VERONEN, L.J., BEST, C.L., & VON, J.M. (1987). Criminal victimization: Lifetime prevalence, reporting to police, and psychological impact. *Crime and Delinquency, 33*(4), 479–489.

KNUDTEN, R.D., MEADE, A., KNUDTEN, M., & DOERNER, W. (1976). The victim in the administration of criminal justice: Problems and perceptions. In W.F. McDonald (Eds.), *Criminal justice and the victim* (pp. 115–146). Beverly Hills, CA: Sage.

Law Enforcement Assistance Administration. (1982). *Criminal victimization in the United States: 1980.* Washington, DC: U.S. Government Printing Office.

MAGUIRE, M., & CORBETT, C. (1987). *The effects of crime and the work of victims support schemes.* Aldershot, UK: Gower.

MAGUIRE, K., & PASTORE, A.L. (Eds.). (1996). *Sourcebook of criminal justice statistics,* Washington, DC: U.S. Government Printing Office.

MAYHEW, P., ELLIOTT, D., & DOWDS, L. (1989). *The 1988 British Crime Survey.* London: Office Her Majesty's Stationary.

MAWBY, R.I., & WALKLATE, S. (1994). *Critical victimology.* Thousand Oaks, CA: Sage.

McDONALD, W.F. (1976). *Criminal justice and the victim.* Beverly Hills, CA: Sage.

Office of Community Oriented Policing Services. (1997a). *About COPS.* Washington, DC: U.S. Department of Justice.

Office of Community Oriented Policing Services. (1997b). *COPS office report: 100,000 officers and community policing across the nation.* Washington, DC: U.S. Department of Justice.

Office of Community Oriented Policing Services. (1997c). *Community policing success story no. 6: Lowell, Massachusetts.* Washington, DC: U.S. Department of Justice.

Office of Justice Programs. (1986). *Four years later: A report on the president's task force on victims of crime.* Washington, DC: Department of Justice.

Office of the Attorney General. (1995). *Attorney general guidelines for victim and witness assistance.* Washington, DC: U.S. Department of Justice.

President's Task Force on Victims of Crime. (1982). *Final report.* Washington, DC: U.S. Government Printing Office.

REISS, A. (1971). *The police and the public.* New Haven, CT: Yale University Press.

RUBACK, R.B., GREENBERG, M.S., & WESTCOTT, D.R. (1984). Social influence and crime victim decision making. *Journal of Social Issues, 40*(1), 51–76.

SALES, E., BAUM, M., & SHORE, B. (1984). Victim readjustment following assault. *Journal of Social Issues, 40*(1), 117–136.

SCHNEIDER, A.L., BURCART, J.M., & WILSON, L.A. (1976). The role of attitudes in the decision to report crimes to the police. In W.F. McDonald (Ed.), *Criminal justice and the victim* (pp. 89–113). Beverly Hills, CA: Sage.

SCHWIND, H.D. (1984). Investigation of nonreported offenses: Distribution of criminal offenses not known to authorities. In R. Block (Ed.), *Victimization and fear of crime: World perspectives* (pp. 65–74). Washington, DC: U.S. Government Printing Office.

SHAPLAND, J., WILLMORE, J., & DUFF, P. (1985). *Victims in the criminal justice system.* Aldershot, UK: Gower.

SKOGAN, W.G. (1984). Reporting crimes to the police: The status of world research. *Journal of Research in Crime and Delinquency, 21*(2), 113–137.

SPARKS, R.F., GENN, H.G., & DODD, D.J. (1977). *Surveying victims: Measurement of criminal victimization, perceptions of crime, and attitudes to criminal justice.* London: Wiley.

SPELMAN, W., & BROWN, D.K. (1981). *Calling the police: Citizen reporting of serious crime.* Washington, DC: Police Executive Research Forum.

VAN KIRK, M. (1978). *Response time analysis: Executive summary.* Washington, DC: Law Enforcement Assistance Administration.

WALLER, I. (1990). The police: First in aid? In A.J. Lurigio, W.G. Skogan, & R.C. Davis (Eds.), *Victims of crime: Problems, policies and programs* (pp. 139–156). Newbury Park, CA: Sage.

WALLER, I., & OKIHIRO, N. (1978). *Burglary: The victim and the public.* Toronto: University of Toronto Press.

WEBB, V.J., & MARSHALL, I.H. (1989). Response to criminal victimization by older Americans. *Criminal Justice and Behavior, 16*(2), 239–259.

WEISHEIT, R.A., FALCONE, D.N., & WELLS, L.E. (1994). *Rural crime and rural policing.* Washington, DC: U.S. Department of Justice.

WORTMAN, C.B. (1983). Coping with victimization: Conclusions and implications for future research. *Journal of Social Issues, 39*(2), 195–221.

ZAWITZ, M.W., KLAUS, P.A., BACHMAN, R., BASTIAN, L., DeBERRY, M.W., RAND, M.R., & TAYLOR, B.M. (1993). *Highlights from 20 years of surveying crime victims: The National Crime Victimization Survey, 1973-92.* Washington, DC: U.S. Department of Justice.

4

Policing and Sexual Assault

Strategies for Successful Victim Interviews

Tracy Woodard Meyers

❖

INTRODUCTION

Individuals who experience sexual assault are victims of horrifying and traumatic crimes. Often fearing for their lives, they seldom escape physically and emotionally unscathed. While physical injuries may range from bruising to death, it is not unusual for victims to experience both short- and long-term emotional symptoms. Some researchers suggest that as many as 50% of women who have been raped suffer from post-traumatic stress disorder (Rothbaum, Foa, Riggs, Murdock, & Walsh, 1992).

Although the actual experience of sexual assault proves to be correlated with traumatic stress symptoms, it is not the only event that has the ability to impact victims' emotional states. The manner in which victims are treated after sexual assaults also impacts their emotional well-being and recovery process. Pennebaker, Kiecolt-Glaser, and Glaser (1988) noted that disclosing traumatic events has favorable psychological and physical health consequences for victims. The reactions of the persons to whom victims disclose their traumas, however, are instrumental in determining whether the process is beneficial or harmful. When victims are treated in an empathic and supportive manner, it helps with the healing process and proves to be psychologically beneficial (Madigan & Gamble, 1991; Fairstein, 1993). When victims are treated in an insensitive manner, they often experience an increase in feelings of powerlessness, shame, and guilt, experiencing a secondary victimization (Madigan & Gamble, 1991).

When treated insensitively, victims are not the only ones who stand to suffer. The outcomes of the cases may be negatively influenced as well. Police officers are generally the first professionals to see and interview sexual assault victims after their traumatic ordeals. The first responding officer's duties and responsibilities are crucial in determining the direction of the investigation. It is therefore imperative that these professionals understand the impact their interactions have on victims. Accusations of police insensitivity toward victims are all too common (Powers, 1996). Victims who are treated poorly tend to be less willing to cooperate with the police (Burgess & Hazelwood, 1995). Conversely, when victims are treated in an empathic and supportive manner, they are more likely to cooperate with law enforcement officials and are often able to recount more information (Ullman, 1996). Unfortunately, first responding police officers often do not have special training in sexual assault issues. Powers (1996, p. 1) contends police officer insensitivity is "merely a reflection of the average uniform officer's and detective's lack of specialized training and education in the field of victimology." Therefore, adequately educating police officers about sexual assault issues and training them in appropriate victim interviewing techniques should increase the likelihood of police officers and victims having a positive working relationship while decreasing the chances of revictimization. The purpose of this chapter is to provide professionals with the information and skills they need in order to interact successfully with sexual assault victims.

DEFINING SEXUAL ASSAULT

Prior to recent reforms, rape was defined as "carnal knowledge (penile-vaginal penetration only) of a female forcibly and against her will" (Bienen, 1981, p. 174). Because this definition did not allow for male victims or take into account other types of sexual activity, advocates lobbied to get the laws and definitions modified. A majority of states have revised their rape laws to include gender neutral language and have expanded the definition beyond penile-vaginal intercourse (Roberts, 1997). Many have even replaced the term *rape* with *sexual assault* in order to be more inclusive. For the purposes of this chapter, the definition of sexual assault presented by Hall (1995, p. xiii) will be used. Sexual assault "will include but will not necessarily be limited to, actual or attempted vaginal or anal intercourse, fellatio, cunnilingus, and penetration by a foreign object or digit, performed against the will of the victim and regardless of the victim's response to the attack."

SCOPE OF THE PROBLEM

Sexual assault is a prevalent crime in the United States. As many as 307,000 women in 1996 (U.S. Department of Justice, 1997) and 32,130 males in 1995 (U. S. Department of Justice, 1996) were reported to be victims of sexual assault. A 1992 national survey suggested that one out of every eight women has been a victim of rape (Kilpatrick, Edmunds, & Seymour, 1992). Although most think that sexual assaults are perpetrated by strangers, the reality is that approximately 80% of all sexual assailants are known to their victims (Hall, 1995). Despite its pervasiveness and high probability of identifying the offenders, sexual assault is the most underreported crime in the United States (George, Winfield, & Blazer, 1992; Koss, 1988; Roberts, 1997). Rochman (1991) contends that 50% of women who have been raped have never told friends or family members about the assault and only one in 50 women reports the crime to police. Other research suggests nonreporting rates as high as 88% (Kilpatrick et al; 1992), 84% (Garcia & Henderson, 1999), and 77%

(Wyatt, 1992). Rates of underreporting are believed to be even higher for men (Rochman, 1991; Scarce, 1997).

A number of factors have been recorded as deterrents for reporting sexual assault to the police. Some victims do not recognize that they are victims of a crime and thus do not report (Neville & Pugh, 1997). Some may believe it is a private or personal matter or they fear reprisal from the assailant. Others may be in a state of crisis or suffering from post-traumatic stress symptoms that may impair their ability to report. Many, however, are concerned about social stigmas and the treatment they will receive from the professionals with whom they come in contact. Because of the acceptance of widespread rape myths, victims fear they will not be believed or will be blamed for the rape. Women have reported feeling humiliated, intimidated, and shamed when questioned by law enforcement about their truthfulness, motives, and moral character (Frohmann, 1998). Men report being "too embarrassed" to report and do not want to deal with "being judged" by the police (Rochman, 1991).

Police officers may interact negatively with victims or dismiss cases as "unfounded" because of their personal beliefs concerning sexual assault myths and stereotypes. In a study measuring police officers' perceptions of sexual assault, Campbell and Johnson (1997) found that officers held widely differing beliefs as to what constitutes sexual assault. Although 19% of the sample defined sexual assault consistently with state law, 50% of the officers surveyed ascribed to stereotypical ideas. When officers were asked to write in their own words how they defined sexual assault, those with stereotypical ideas contributed statements full of desecrating myths. For example:

> "What if someone has had sex 20 or 30 times over a 3-month period. Then one night they say "no." Should this be rape? I don't think so."
>
> "I do not believe that consensual sex that is later regretted and reported as nonconsensual sex is rape. The 'man in the bushes' is a much bigger threat to the community than the date that got carried away."
>
> "Sometimes a guy can't stop himself. He gets egged on by the girl. Rape must involve force—and that's really rare."
>
> "Men taking what women really want at that moment but decide they didn't the next morning when they sober up."
>
> "Rape is just rough sex that a girl changed her mind about later on. Technically, rape is a sex act done by the use of force, but so many girls are into being forced, that you can't tell the difference and you wouldn't want to convict an innocent guy."
>
> "I can tell you what rape isn't: an individual going on a date and getting into foreplay and then finally consenting after being verbally persuaded" (p.268).

Despite statistics that suggest victims are more likely to be assaulted by someone they know, a number of officers express doubts concerning the authenticity of acquaintance sexual assaults (Campbell & Johnson, 1997). In their studies, Fairstein (1993) and Kerstetter (1990) found that stranger rapes were investigated more thoroughly and less likely to be closed as "unfounded" than nonstranger assaults. Simon (1996) suggests the criminal justice system has a bias for treating nonstranger offenders more leniently.

A number of other sexual assault myths have influenced the criminal justice system's treatment of victims and their complaints. LaFree found that officials are more likely to file felony charges if the incident involved an African American male offender and a white female victim (Kerstetter, 1990). According to Rose and Randall, when the victim is nonwhite, prosecutors are more likely to file less serious charges (Kerstetter, 1990). Neville and Pugh (1997) contribute these biased treatments to racist stereotypes that assert African American men are sexually aggressive and African American women are sexually loose. Other studies suggest that assaults involving penetration, use of a weapon, and physical injuries are more likely to be prosecuted (Campbell, 1998; Kerstetter, 1990; Madigan & Gamble, 1991; Russell, 1990). Galton reported that police officers required victims to have resisted their attackers to the point of being injured before they would consider the case legitimate (Kerstetter, 1990). When victims had bad reputations, were hitchhiking, or drinking prior to the sexual assault, criminal justice officials had a tendency to dismiss cases as unfounded (Roberts, 1997).

Police officers' lack of knowledge concerning traumatic stress reactions also determines how victims are treated and whether cases are closed as unfounded or pursued. After experiencing a traumatic event it is not unusual for victims to experience a wide variety of reactions such as fear, anxiety, depressive symptoms, emotional numbness, confusion, and sharp mood shifts (Resick, 1988; Roberts, 1997). Although victims of traumatic events respond in a number of different ways, some police officers have preconceived notions of how victims of sexual assault "should act." Not understanding traumatic reactions, many officers erroneously use the victims' behaviors as a basis for the legitimacy of the assaults. Madigan and Gamble (1991) and Campbell (1998) found that victims who were visibly upset and showed signs of trauma were considered "good victims" and more likely to have their cases prosecuted. Victims with inconsistencies in their accounts of the assault were perceived as less credible and were more likely to have their cases closed as unfounded (Burgess, 1995).

POST-TRAUMATIC STRESS DISORDER

During the 1970s researchers began studying victims' reactions in response to a variety of traumatic events such as rape, domestic violence, and child abuse, to name a few. The focus at this time was on specific types of psychosocial stressors and the different symptoms they elicited. A number of traumatic stress syndromes were identified in the literature including the rape-trauma syndrome by Burgess and Holmstrom, the post-Vietnam syndrome by Shatan, and the Holocaust survivor syndrome by Niederland (Figley, 1988). It was not until 1980, however, with the publication of the *Diagnostic and Statistical Manual of Mental Disorders, 3rd Edition* (DSM-III) (American Psychiatric Association, 1987) that scholars acknowledged that persons who experienced traumatic events, regardless of its nature, share common symptoms. The DSM-III included the diagnosis of post-traumatic stress disorder (PTSD). For the first time, trauma victims were diagnosed as suffering from a psychiatric disorder and treatment was thus prescribed. Since that time, the DSM-III has undergone two revisions with the symptom criteria for PTSD being modified both times. The most recent *Diagnostic and Statistical Manual of Mental Disorders, 4th Edition* (APA, 1994) includes the common symptoms of both adult and children for the diagnosis of PTSD. PTSD is characterized by avoidance symptoms, reexperiencing the traumatic event, and increased arousal. With the recognition of PTSD and its acceptance as a disorder, the number of professionals working with trauma victims has grown as has the accumulation of empirical research that validates its existence.

PREVALENCE OF TRAUMATIC STRESS REACTIONS IN SEXUAL ASSAULT VICTIMS

Studies have been conducted on a variety of victim populations to determine the prevalence of traumatic stress symptoms and PTSD. Populations that have been studied include child abuse victims (Goodwin, 1988), sexual abuse and rape victims (Burgess & Holmstrom, 1974; Kilpatrick & Resnick, 1993; Roth, Wayland, & Woolsey, 1990), war and combat veterans (Blake, Keane, Mora, Taylor, & Lyons, 1990; Solomon, 1989), survivors of the Holocaust (Danieli, 1984), natural disaster victims (Galante & Foa, 1986), technological disaster victims (Bartone & Wright, 1990), hostage victims (van der Ploeg & Kleijn, 1989), witnesses of homicides (Waigandt & Phelps, 1990), spousal abuse victims (McCormack, Burgess, & Hartman, 1988), emergency workers (McCammon, Durham, Allison, & Williamson, 1988), and police officers (Gersons, 1989). Research findings confirm that traumatic stress symptoms as well as PTSD are common among all of these populations following exposure to traumatic events.

Kilpatrick and Resnick's (1993) study of women who had experienced aggravated assault reported that 39% of the sample developed PTSD. In those women who were raped, 35% developed PTSD. Rothbaum, Foa, Riggs, Murdock, and Walsh (1992) found all the rape victims in their study had the symptom criteria for PTSD immediately following the assault. Rape trauma syndrome (RTS), a form of post-traumatic stress disorder, was coined by Ann Wolbert Burgess and Lynda Holmstrom in 1974 to describe "an acute phase and long-term reorganization process that occurs as a result of sexual assault" (Burgess & Hazelwood, 1995, p. 27). According to Burgess and Holmstrom (1974), all sexual assault victims suffer from some degree of RTS. Gidycz and Koss (1991) noted that immediately after the assault, victims most commonly report feelings of anxiety or fearfulness. Other immediate reactions commonly reported by victims are physical symptoms, fear of dying, guilt, shame, anger, humiliation, sleep pattern disturbances, and flashbacks (Burgess & Hazelwood, 1995).

INTERVIEWING SEXUAL ASSAULT VICTIMS

Based on the findings from the review of literature and informal interviews with sexual assault victim advocates, a key factor in determining the likelihood that police officers and sexual assault victims will have a positive working relationship is a successful interview process. The manner in which law enforcement officials conduct themselves during the interview process is crucial in determining the direction of the investigation as well as the victims' emotional states and recovery processes. When treated in an empathic and supportive manner, victims tend to be cooperative, recall more information, and reap psychological benefits (Fairstein, 1993; Madigan & Gamble, 1991; Ullman, 1996). When victims are treated insensitively, they are less cooperative and often experience a secondary victimization (Madigan & Gamble, 1991). A successful interview would be one in which officials retrieve valuable case information while victims gain the psychological benefits that result from disclosing traumatic events in a supportive atmosphere.

There are a number of factors that impede officers' ability to conduct successful interviews with sexual assault victims. One factor is officers' biased beliefs and attitudes that are based on erroneous sexual assault myths and stereotypes. In order to conduct a successful interview, police/law enforcement officials need to be aware of how myths and stereotypes influence their personal belief systems and sway their perceptions and judgments. Another influencing factor is the lack of knowledge and understanding officers have concerning typical traumatic stress reactions. Officers should be educated about traumatic stress reactions so they can realize that the legitimacy of an assault should not be based on victims' behaviors. Erroneous beliefs about how victims "should act" influence officers' judgment and negatively impact the interactions between officers and the victims. Finally, the lack of sensitive interviewing skills that officers display while questioning victims also hampers the inquiry process. Officers should refrain from the use of interrogation style interviewing techniques when soliciting information from victims of crimes. Although interrogation techniques may be useful with perpetrators, they have detrimental effects on victims. An empathic and supportive interview style needs to be incorporated when working with victims (see Chapter 6 for interviewing strategies).

STRATEGIES FOR POSITIVE WORKING RELATIONSHIPS

There are a number of interventions that can be utilized in order to increase the likelihood of police officers and sexual assault victims having a positive working relationship. One way to ensure a positive relationship is having a successful interview process. In order to conduct a successful interview, law enforcement officials should have an awareness of how sexual assault myths influence their personal beliefs, an understanding of traumatic stress reactions, and sensitive interviewing techniques. The following information will help officers conduct successful inquiries.

Becoming Aware of Sexual Assault Stereotypes and Personal Biases

Not only do police officers need to be aware of sexual assault stereotypes and myths, they also should be aware of their own personal beliefs and attitudes concerning sexual assault. Brownmiller (1975) affirms that despite officers' knowledge of the law, the male police mentality is often identical to the stereotypical views of sexual assault that are shared by the male culture. These bias attitudes cloud officers' judgment and are often communicated to the victims. In order to avoid subscribing to such biased beliefs and attitudes, police/law enforcement officials need to be aware of sexual assault stereotypes and myths. (See Appendix A for a list of common misconceptions concerning sexual assault as presented by Hall, 1995.)

Understanding Victims' Traumatic Stress Reactions

Stress can be defined as the process by which events or stressors tax the coping abilities or pose a threat to a person (Davidson, Fleming, & Baum, 1987). According to Caplan (1964), individuals in noncrisis states maintain a level of emotional balance, in which stress is successfully handled by problem-solving skills. When confronted with stressors, a person must find ways to deal with the demands, ways to adapt. A response thus occurs that is designed to remove the stressor and its effects until the episode subsides. As this process progresses, psychological and physiological changes occur in order to carry out this task. Cannon (1929) referred to this as the "fight-or-flight" response.

During the fight-or-flight response, physiological and mental conditions of the body change to allow the person to fight or run from the stressor. At this point the muscles tighten, the pupils dilate to take in more light, breathing rate increases, heart rate increases, blood pressure increases, and cholesterol and triglyceride levels increase rapidly in the blood. The fat cells—the fuel for muscles—appear in abundance and the protein and antibody levels in the blood increase dramatically. All of these changes occur to enable the person to physically respond to the stressor at hand.

Just as the physiological changes occur to prepare the body to fight or run from the stressor, so does the person's emotional and cognitive functioning. As individuals experience increasing levels of stress, they begin to lose their mental efficiency. The ability to remember new information begins to diminish and concentration becomes difficult. Some people experience a flow of ideas that appear to be on high speed, and they are unable to select one idea on which to focus. Perceptions of problems change so that minor problems appear to be major and one's ability to be mentally flexible becomes more rigid. Therefore, the ability to solve problems and determine alternative solutions is impaired.

Emotional changes also occur at this time. When faced with stressors, individuals tend to become more rigid, which results in the loss of an effective coping mechanism—humor. People lose their ability to laugh at themselves and the situation. People also lose their ability to trust others and often isolate themselves by withdrawing from their support. When people lose their defense mechanisms and support they tend to replace them with fantasies and wishful thinking. These cognitive and emotional changes occur to minimize the probability of emotional overload. The changes provide a psychological barrier that enables persons to function with a minimal amount of distracting emotional energy. Therefore, they are allowed to concentrate on the task at hand, adapting to the stressor.

After the situation is resolved, and the threat no longer exists, additional changes or aftereffects may occur. These changes further assist in the adaptation to the stressor and return to a prestressor state. Unfortunately, attempts to manipulate or adapt to the stressors are not always successful. The intensity and persistence of the stressor and the person's ability to cope determine whether the proper accommodations can occur. If not, negative consequences are likely to develop. These consequences involve both physiological and psychological symptoms ranging from mild to severe.

Within the past decade, professionals in the mental health field have discovered that the intensity and persistence of a traumatic stressor—"an event outside the realm of normal human experience" (Figley, 1995, p. 3)—often leaves individuals so overwhelmed and their coping abilities so taxed, that they are not able to adapt as noted earlier (Figley, 1985; van der Kolk, 1987; Wilson, 1989). When a situation is such that coping abilities are inadequate, an individual may experience a crisis reaction. A crisis is defined as an imbalance between the perceived difficulty and significance of a threatening situation and the coping resources available to an individual (Caplan, 1964).

According to Caplan (1964), a crisis reaction is characterized by specified developmental sequences that begin with a threatening situation. Once individuals are exposed to a threatening situation, feelings of anxiety appear and problem-solving strategies are called upon to restore emotional balance. When the problem-solving strategies fail, the individuals' feelings of tension are further exacerbated. At this point, functioning may become disorganized and trial-and-error attempts at coping may appear. Emergency and new coping responses emerge in an attempt to reduce anxiety. If these emergency measures fail, anxiety is further heightened, the individuals reach their breaking point, and personality disorganization follows.

When individuals exceed their physical limitations of responding and their adaptation capabilities are exhausted, post-traumatic stress symptoms (physiological and psychological responses) develop (Davidson et al., 1987). Although the severity and longevity of the symptoms vary among individuals, when these symptoms linger too long or have a delayed onset, they are considered pathological, and a post-traumatic stress disorder (PTSD) (APA, 1994).

The central PTSD symptoms that can result in response to a traumatic event include (1) the persistent avoidance of stimuli associated with the trauma; (2) the persistent reexperiencing of the traumatic event; and (3) the persistent symptoms of increased arousal (APA, 1994). During and immediately following a traumatic event, a person may experience a numbing or avoidance reaction (Everly, 1995; Figley, 1985; Horowitz & Kaltreider, 1995). When individuals experience avoidance reactions, symptoms such as "inability to recall important aspects of the trauma," "restricted range of affect," and "efforts to avoid thoughts, feelings, or conversations associated with the trauma" appear (APA, 1994, p. 428). Avoidance reactions protect the mind from the overwhelming pain of the trauma.

Unfortunately, these avoidance reactions coexist with periods of reexperiencing the trauma through uncontrollable and emotionally distressing intrusive images (Everly, 1995; Figley, 1983, 1985; Horowitz & Kaltreider, 1995; Matsakis, 1994). The reexperiencing of the trauma alternates with avoidant states and the two are thought to occur in cycles (Everly, 1995; Figley, 1983; Green, Wilson, & Lindy, 1985; Horowitz & Kaltreider, 1995; Matsakis, 1994). The reexperiencing of the event may be seen in the form of nightmares and flashbacks (the sudden vivid recollection of the event accompanied by strong emotions). Physiological reactivity can also occur including such symptoms as sweating, rapid heartbeat, nausea, dizziness, dry mouth, hot flashes, chills, frequent urination, trouble swallowing, and diarrhea (Matsakis, 1994).

Another category of PTSD symptoms that occur in response to a traumatic stressor is increased arousal or hyperarousal (APA, 1994; Everly, 1995; Figley, 1983, 1985; Horowitz & Kaltreider, 1995; Matsakis, 1994). During the actual traumatic event, the fight-or-flight reaction occurs, changing the body's physiological and emotional conditions. Because the body readies itself to run away or fight the stressor, large quantities of adrenaline are released in the body, putting it in a hyperalert stage. When individuals experience increased arousal symptoms, it is because the brain tells the body that the threat is still imminent so the body continues to release high levels of adrenaline into the system. Hyperarousal symptoms are often seen in individuals as irritability, outbursts of anger, difficulty concentrating, difficulty falling or staying asleep, and an exaggerated startle response (APA, 1994).

As previously stated, it is normal to experience post-traumatic stress symptoms after being exposed to a traumatic event. However, if the symptoms are prolonged, blocked, or exceed a tolerable quality, they are pathological and the person is said to suffer from PTSD (Zilberg, Weiss, & Horowitz, 1982). In order for individuals to be diagnosed as suffering from PTSD, they must first be exposed to a traumatic event. The DSM-IV (APA, 1994) defines a traumatic event as:

> the direct personal experience of an event that involves actual or threatened death or serious injury, or other threat to one's physical integrity; or witnessing an event that involves death, injury, or a threat to the physical integrity of another person; or learning about unexpected or violent death, serious harm, or threat of death or injury experienced by a family member or other close associate (p. 424).

Second, the person's response has to involve intense fear, helplessness, or horror. Finally, the symptoms must be present for more than one month or occur at least six months after the traumatic event has passed. As a result, the symptoms must cause significant distress or impairment in the person's social, occupational, or other important areas of functioning.

RAPE TRAUMA SYNDROME

Rape trauma syndrome (RTS) is a predictable traumatic stress reaction victims of sexual assault typically display following an assault (Burgess & Holmstrom, 1974). It is used to describe behavioral, somatic, and psychological reactions that occur as a result of a sexual assault or attempted sexual assault. The reactions manifest in two stages, an acute phase and a long-term reorganization process. The acute phase involves victims displaying general stress response symptoms such as fluctuating mood swings, crying, panic, fear, anxiety, laughing, or shock. The acute phase can last from several hours to several weeks. It is likely that victims will experience a variety of emotional and physical reactions during this phase.

The reorganization process follows the acute phase and is described as a time in which victims try to reorganize their lives in order to regain a sense of control. During this stage it is not unusual for victims to engage in lifestyle changes (i.e., new residence or new telephone number) or suffer from long-term chronic disturbances (i.e., nightmares, phobias, and sexual problems). This phase can last for six months to a year or longer. As reported earlier, it is thought that all sexual assault victims suffer from some degree of rape trauma syndrome (Burgess & Holmstrom, 1974).

The Acute Phase

During the acute phase, it is not unusual for victims to display a wide range of emotions. After experiencing a sexual assault some victims are capable of openly displaying their emotions, whereas others are not. If victims have an "expressive style," their feelings can be observed as they demonstrate them through body language or facial expressions (Burgess & Hazelwood, 1995). Typical emotional responses such as dissociative reactions, suicide ideation, anxiety, fear, difficulty concentrating, depression, and anger are thereby easy to discern. Other victims' emotions may not be as easily recognized. Victims who subscribe to a "guarded style" often appear calm and composed. Some investigators may mistakenly perceive the lack of emotional reactions to mean the sexual assault did not occur. To the contrary, a number of victims appear composed because they are experiencing shock, a sense of disbelief, denial, dissociative reactions, or physical exhaustion. In this situation, the victims' true feelings are masked or hidden.

Victims often suffer from a number of physical symptoms following a sexual assault. It is not unusual for them to experience soreness, skeletal muscle tension, gastrointestinal irritability, genitourinary disturbances, and vaginal pain (Block, 1990; Burgess & Hazelwood, 1995). In addition, victims also experience sleep pattern disturbances, eating pattern disturbances, social problems, and sexual problems (Foa & Rothbaum, 1998).

Long-Term Reorganization Process

Sexual assault disrupts the normal social lifestyles of its victims. The reconstruction of lifestyles and long-term chronic disturbances are typical characteristics of the reorganization process. While trying to regain control, it is not uncommon for victims to experience excessive fears and phobias. Maladaptive fears and phobias may result in victims staying at home, only leaving their houses when accompanied by friends, moving, changing telephone numbers, or no longer being able to work or attend school (Burgess & Hazelwood, 1995). It is also common for victims to experience a decrease in sexual satisfaction, fear of sex, and a decrease in sexual arousal or desire (Foa & Rothbaum, 1998). Victims in this phase may be more concerned with regaining control of their lives and feeling safe than cooperating with criminal investigations.

THE SUCCESSFUL INTERVIEW

In order to conduct a successful interview, police officers should adhere to the following guidelines:

When and Where to Interview

When possible, allow the victim to decide when and where the interview will take place. This helps provide the victim with a sense of control. An interview should never be conducted while the victim is undergoing a medical sexual assault exam or is in an environment in which she or he feels unsafe. A quiet setting free from distractions is preferred.

Victims should be allowed to choose who is present during the interview. Having experienced a traumatic event, the victim is often feeling anxiety, fear, and a variety of other emotions. A support person can help in comforting the victim and provide her or him with the compassion she or he needs to get through the process.

It is important for the victim to know that her or his immediate needs are understood and attended to by the officer. The officer should ask the victim if she or he needs something to drink, if the room is too cold or hot, if the lighting is okay, and so on. It is important to attend to the victim's physical needs.

Who Should Conduct the Interview

The best person to conduct the sexual assault interview is someone who has been trained in interviewing victims and is sensitive, nonjudgmental, and supportive. Although some female victims may prefer to be interviewed by a female officer, it is best if the victim is asked who she or he prefers as an interviewer. In terms of child victims, it again appears that the gender does not matter as much as the training and the sensitivity of the interviewer.

Assess Yourself

An ongoing self-analysis on the part of the officer is mandatory during the interview process. It is essential that the officer accurately sense all the spoken and unspoken cues, messages, and behaviors he or she is expressing to the victim. Officers must be aware that their own values and attitudes are often expressed both verbally and nonverbally. Verbal statements, body posture, body movement, gestures, facial expressions, vocal pitch, eye contact, movement of the arms and legs, and other body indicators all need to be assessed for accurate and congruent messages. The goal is to express an empathic understanding to the victim through both verbal and nonverbal communication.

Building Rapport

Establishing a rapport with the victim is only one of many reasons to employ empathy. Empathy involves not only understanding and being sensitive to how the victim feels but also expressing this understanding to the victim. When expressing empathy, officers need to be aware of their facial expressions and body postures. To convey a sense of involvement, concern, commitment, and trust, officers need to use effective attending skills. Examples of attending skills are as follows: nodding, keeping eye contact, smiling, showing appropriate seriousness of expression, leaning forward, keeping an open stance, and sitting or standing close to the victim without invading her or his space. In addition, officers can verbally communicate empathic understanding by translating what they think the victim is feeling into words. The following are leading phrases that can be used to begin an empathic statement: "My impression is that . . ," "It appears to me that. . . ," "You seem to be . . . ," "You look like you . . . ," and "I'm getting the message that . . . ." The central issue in empathic understanding is to be in tune with the victim's feelings and concerns and express them back to the victim.

Initial Introductions

The officer should introduce him- or herself in a professional and sincere manner. The victim should initially be addressed by her or his last name preceded by the appropriate title (i.e., Mrs., Miss, Ms., or Mr.). During the introductions, the officer needs to express remorse that the assault occurred and ensure the victim that she or he did not deserve the violation. The victim needs to be told clearly that the assault was not her or his fault. The victim needs to be reassured that she or he is safe and everything possible will be done to maintain that safety.

The officer should explain the methods and stages of the investigation to the victim. The purpose of the interview as well as the role of the officer should be clarified for the victim. The officer should request the victim's feedback to ensure that she or he understands the procedures and to help the victim feel like a "partner" in the process. The victim should be informed that if she or he does not understand a question, she or he can ask for clarification. In addition, the victim needs to know that she or he can take as many breaks as are needed, can leave whenever she or he desires, and may speak with a victim advocate at any point during the process.

Conducting the Interview

Officers should use open-ended questions to elicit information from the victim. The victim should be asked to describe the assault and be allowed to provide information in her or his own words. For example, an officer may say to the victim, "Please describe the assault." The victim should be allowed to tell what happened at her or his own pace and without interruptions.

Clarifying questions may be asked to establish one or more facts after the victim has told her or his story. Clarification should be used when there is a question about what the victim means. The information that needs clarifying should be repeated using the victim's words. The victim should be encouraged to interrupt to include a fact or correct a mistake.

Officers need to understand that victims may become uncomfortable when disclosing personal and highly embarrassing information. Sometimes, due to discomfort, victims will omit certain details or information. Throughout the interview, officers should pay close attention to the victim's body language and other nonverbal responses in order to discern discomfort. When the victim becomes uncomfortable, the officer needs to acknowledge that the questions are embarrassing and normalize the discomfort the victim is feeling. Reassuring the victim that she or he is doing a good job and reexplaining why the information is needed will encourage the victim to continue. Reassuring comments should be provided throughout the interview.

Concluding the Interview

After the disclosure, the victim may look to the officer for a reaction and assessment of guilt. The officer should end the interview by complimenting the victim on her or his ability to survive the attack and by thanking the victim for the time she or he has taken in assisting in the investigation. The victim should be advised of the next steps in the investigative process and should be given a number to call to be kept apprised of the investigation. If not already provided, the victim should be provided with referral information on support services (see Chapter 8).

What to Avoid During an Interview

Officers should avoid putting victims through lengthy interviews especially if they have recently experienced an assault. Victims experience a number of emotions in the aftermath of the traumatic event making it difficult for them to concentrate and recall information. In addition, some victims may have been provided with medication to decrease the chances of pregnancy, the result of which often renders them physically ill.

Officers need to remember that individuals who experience sexual assault are the victims of a crime and should be treated as such. A victim should not be asked to take a polygraph test nor should she or he be questioned in an interrogating fashion. Questions concerning the victim's sexual history should be avoided since this type of information has nothing to do with the crime of sexual assault. Under no circumstances should the victim be blamed for the attack even if she or he was engaged in high-risk behavior. For example, victims should never be asked why they were out late, or why they were wearing certain clothing.

CONCLUSION

Sexual assault is a prevalent crime in the United States that leaves its victims suffering from devastating post-traumatic stress reactions. Police officers have the capabilities of not only determining the direction of the investigation but the victims' emotional states and recovery processes as well. All too often, responding police officers and detectives do not have the specialized training or education needed to successfully work with assault victims. Victims end up being revictimized by the interview process and cases are closed as "unfounded" due to the lack of cooperation. In order to prevent such atrocities, officers need to be aware of how myths and stereotypes influence their personal belief systems and influence their perceptions of the victim and the case. Officers also need to understand traumatic stress reactions so they realize that the legitimacy of a case should not be based on the victim's behaviors. Finally, officers should develop an empathic and supportive interview style when working with victims.

APPENDIX A

The following is a list of common misconceptions concerning sexual assault as presented by Hall (1995):

Myth: *Sexual assault has nothing to do with sex (or) sexual assault is a crime of sexual passion.*

Rape is a form of sexual assault and is an act of aggression. It is a crime of violence that uses sexuality as a weapon. The rapist uses sex to express anger, to control, degrade, humiliate, and hurt the victim. Sexual assault is a sexual act (including forced intercourse, sodomy, cunnilingus, fellatio, and penetration by a digit or foreign object) done against the will of the victim. Rape is a sexual behavior done primarily to satisfy nonsexual desires.

Myth: *Rapes usually occur at night in alleys, parks, and the like.*

A high percentage of rapes occur in the victim's home, or at the home of someone the victim trusted; the next most common place of assault is in a car.

Myth: *Rape usually involves a black assailant and a white victim.*

Numerous studies have shown rape to be primarily intraracial, not interracial. Victims and assailants are most frequently of the same race.

Myth: *If a woman leads a man on, or allows him to spend a great deal of money on her, or changes her mind after having commenced foreplay, the man has a right to sex.*

Sexual assault is a forced sexual act without consent. It is a crime regardless of the previous actions of the victim or the assailant.

Myth: *If there was no semen present, there was no rape.*

Studies of convicted rapists have found that between 34% and 58% experienced sexual dysfunction in the form of impotence, premature ejaculation, or retarded ejaculation, thus accounting for a lack of semen present.

<u>Myth:</u> *The sexual assailant is most often a stranger.*

The most common estimate is that approximately 80% of all sexual assailants are known to the victim.

<u>Myth:</u> *A rapist is a psychopath and looks like one.*

Rapists come from all walks of life, frequently have families, and include clergy, policemen, teachers, and others who are generally respected as "model citizens." Most often the rapist is someone known to, and trusted by, the victim.

<u>Myth:</u> *The rapist is most frequently a "sex starved" pervert.*

The majority of convicted rapists had regular sexual outlets at the time of their offenses. They did not rape out of sexual frustration, but for the emotional gratification they received from the act of sexual violence.

<u>Myth:</u> *A rapist is a man who cannot control his sexual desires.*

Rape is most often a premeditated crime. It is an act of aggression and sexual violence, not an expression of sexual desire.

<u>Myth:</u> *A husband cannot be found guilty of raping his wife.*

All 50 states have removed the automatic "spousal exemption" from their laws on sexual assault. A husband in the United States can be tried for the rape of his wife.

<u>Myth:</u> *The victim was "asking for it" by the manner in which she dressed, by flirting, or by where she was walking or spending her time (such as a bar or the assailant's dwelling).*

Rape is an act of violence that hurts the victim physically and psychologically. The attitude that the victim "asked for it" takes the responsibility for the attack away from the assailant, shifting it to the victim.

<u>Myth:</u> *Rape only happens to certain kinds of women.*

Rape happens to women of all ages, races, socioeconomic status, ethnic, and religious groups.

<u>Myth:</u> *No one would rape an "unattractive" woman; rape only happens to young, "sexy" women.*

Victims are chosen on the basis of their vulnerability, not their physical appearance.

<u>Myth:</u> *No woman can be raped against her will; it is physically impossible to rape a woman who does not want to be raped.*

Rape is most frequently committed through the use of force or threat of force. It is a physical assault using sex. Any person can be physically overcome by a larger and stronger assailant.

<u>Myth:</u> *A male cannot be raped.*

The rape of males is believed to be even more underreported than that of females. Although both male children and male adults have been raped by females, they are most frequently raped by other heterosexual males.

<u>Myth:</u> *Women enjoy being raped.*

Rape is not an act of "rough sex." Rape by its very nature is an unwanted act of violence against both the body and mind of the victim. It violates and destroys the victim's normal perception of and assumptions about her world; highly valued beliefs of trust and safety are shattered.

Myth: *Women frequently have an orgasm while being raped.*

Although it is theoretically possible for a woman or a man to experience an orgasm while being raped, it is a very rare occurrence. An orgasm under those conditions is a physiological response; a result of fear/adrenaline and direct stimulation through cunnilingus, fellatio, or manual stimulation. It is not an indication of sexual arousal (pp. 7–9).

Reprinted with permission from Rape in America: A reference book. Copyright 1995 ABC-CLIO.

REFERENCES

AMERICAN PSYCHIATRIC ASSOCIATION. (1987). *Diagnostic and statistical manual of mental disorders* (3rd ed.). Washington, DC: Author.

AMERICAN PSYCHIATRIC ASSOCIATION. (1994). *Diagnostic and statistical manual of mental disorders* (4th ed.). Washington, DC: Author.

BARTONE, P., & WRIGHT, K. (1990). Grief and group recovery following a military disaster. *Journal of Traumatic Stress, 3*(4), 523–539.

BIENEN, L.B. (1981). Rape III–National developments in rape reform legislation. *Women's Rights Law Reporter,* 171–213.

BLAKE, D., KEANE, T., MORA, K., TAYLOR, K., & LYONS, J. (1990). Prevalence of PTSD symptoms in combat veterans seeking medical treatment. *Journal of Traumatic Stress, 3*(1), 14–28.

BLOCK, A. (1990). Rape trauma syndrome as scientific expert testimony. *Archives of Sexual Behavior, 19*(4), 309–323.

BROWNMILLER, B. (1975). *Against our will: Men, women, and rape.* New York: Simon & Schuster.

BURGESS, A.W. (1995). Public beliefs and attitudes toward rape. In R.R. Hazelwood & A.W. Burgess (Eds.), *Practical aspects of rape investigation: A multidisciplinary approach* (2nd ed., pp. 3–12). Boca Raton, Florida: CRC press. LLC.

BURGESS, A.W., & HAZELWOOD, R.R. (1995). The victim's perspective. In R.R. Hazelwood & A.W. Burgess (Eds.), *Practical aspects of rape investigation: A multidisciplinary approach* (pp. 27–41). Boca Raton, Florida: CRC Press. LLC.

BURGESS, A.W., & HOLMSTROM, S. (1974). Rape trauma syndrome. *American Journal of Psychiatry, 131*(9), 981–985.

CAMPBELL, R. (1998). The community response to rape: Victims' experiences with the legal, medical, and mental health systems. *American Journal of Community Psychology, 26*(3), 355–379.

CAMPBELL, R., & JOHNSON, C.R. (1997). Police officers' perceptions of rape: Is there consistency between state law and individual beliefs? *Journal of Interpersonal Violence, 12*(2), 255–274.

CANNON, W. (1929). *Bodily changes in pain, hunger, fear, and rage: An account of recent researchers into the function of emotional excitement.* New York: Appleton-Century Crofts.

CAPLAN, G. (1964). *Principles of preventive psychiatry.* New York: Basic Books.

DANIELI, Y. (1984). Psychotherapists' participation in the conspiracy of silence about the holocaust. *Psychoanalytic Psychology, 1*(1), 23–42.

DAVIDSON, L., FLEMING, R., & BAUM, A. (1987). Chronic stress, catecholamines, and sleep disturbance at Three Mile Island. *Journal of Human Stress, 13*(2), 75–83.

EVERLY, G. (1995). Psychotraumatology. In G. Everly & J. Lating (Eds.), *Psychotraumatology: Key papers and core concepts in post traumatic stress* (pp. 3–8). New York: Plenum Press.

FAIRSTEIN, L.A. (1993). *Sexual violence: Our war against rape.* New York: William Morrow.

FIGLEY, C. (1983). Catastrophes: An overview of family reactions. In C. Figley & H. McCubbin (Eds.), *Stress and the family: Coping with catastrophe* (pp. 3–20). New York: Brunner/Mazel.

FIGLEY, C. (1985). *Trauma and its wake: The study and treatment of post traumatic stress disorder.* New York: Brunner/Mazel.

FIGLEY, C. (1988). Toward a field of traumatic stress. *Journal of Traumatic Stress, 1*(1), 3–16.

FIGLEY, C. (1995). Compassion fatigue as secondary traumatic stress disorder: An overview. In C. Figley (Ed.), *Compassion fatigue: Coping with secondary traumatic stress disorder in those who treat the traumatized* (pp. 1–20). New York: Brunner/Mazel.

FOA, E., & ROTHBAUM, B. (1998). *Treating the trauma of rape: Cognitive-behavioral therapy for PTSD.* New York: Guilford Press.

FROHMANN, L. (1998). Constituting power in sexual assault cases: Prosecutorial strategies for victim management. *Social Problems, 45*(3), 393–407.

GALANTE, R., & FOA, D. (1986). An epidemiological study of psychic trauma and treatment effectiveness for children after a natural disaster. *Journal of American Academy of Childhood Psychiatry, 25*(3), 357–363.

GARCIA, S., & HENDERSON, M. (1999). Blind reporting of sexual violence. *FBI Law Enforcement Bulletin, 68*(6), 12–16.

GEORGE, L.K., WINFIELD, I., & BLAZER, D.G. (1992). Sociocultural factors in sexual assault: Comparison of two representative samples of women. *Journal of Social Issues, 48*(1), 105–125.

GERSONS, B. (1989). Patterns of post traumatic stress disorder among police officers following shooting incidents: A two-dimensional model and treatment implications. *Journal of Traumatic Stress, 2*(3), 247–257.

GIDYCZ, C.A., & KOSS, M.P. (1991). Predictors of long-term sexual assault trauma among a national sample of victimized college women. *Violence and Victims, 6*(3), 175–190.

GOODWIN, J. (1988). Post traumatic stress symptoms in abused children. *Journal of Traumatic Stress, 1*(4), 475–488.

GREEN, B., WILSON, J., & LINDY, J. (1985). Conceptualizing post traumatic stress disorder: A psychosocial framework. In C. Figley (Ed.), *Trauma and its wake: The study and treatment of post traumatic stress disorder* (pp. 53–69). New York: Brunner/Mazel.

HALL, R. (1995). *Rape in America: A reference handbook.* Santa Barbara, CA: ABC-CLIO, Inc.

HOROWITZ, M., & KALTREIDER, N. (1995). Brief therapy of the stress response syndrome. In G. Everly & J. Lating (Eds.), *Psychotraumatology: Key papers and core concepts in post traumatic stress* (pp. 231–243). New York: Plenum Press.

KERSTETTER, W.A. (1990). Gateway to justice: Police and prosecutorial response to sexual assaults against women. *The Journal of Criminal Law and Criminology, 81*(2), 267–313.

KILPATRICK, D.G., EDMUNDS, C.N., & SEYMOUR, L.J. (1992). *Rape in America: A report to the nation.* Arlington, VA: National Victim Center.

KILPATRICK, D., & RESNICK, H. (1993). Post traumatic stress disorder associated with exposure to criminal victimization in clinical and community populations. In J. Davidson & E. Foa (Eds.), *Post traumatic stress disorder: DSM-IV and beyond* (pp. 113–143). Washington, DC: American Psychiatric Press.

KOSS, M.P. (1988). Hidden rape: Sexual aggression and victimization in a national sample of students in higher education. In A.W. Burgess (Ed.), *Rape and sexual assault* (Vol. 2, pp. 3–25). New York: Garland.

MADIGAN, L., & GAMBLE, N. (1991). *The second rape: Society's continued betrayal of the victim.* New York: Lexington Books.

MATSAKIS, A. (1994). *Post traumatic stress disorder: A complete treatment guide.* Oakland, California: New Harbinger.

MCCAMMON, S., DURHAM, T., ALLISON, J., & WILLIAMSON, J. (1988). Emergency workers' cognitive appraisal and coping with traumatic events. *Journal of Traumatic Stress, 1*(3), 353–361.

MCCORMACK, A., BURGESS, A., & HARTMAN, C. (1988). Familial abuse and post-traumatic stress disorder. *Journal of Traumatic Stress, 1*(2), 231–242.

NEVILLE, H.A., & PUGH, A.O. (1997). General and culture-specific factors influencing African American women's reporting patterns and perceived social support following sexual assault: An exploratory investigation. *Violence Against Women, 3*(4), 361–381.

PENNEBAKER, J.W., KIECOLT-GLASER, J.K., & GLASER, R. (1988). Disclosure of traumas and immune function: Health implications for psychotherapy. *Journal of Consulting and Clinical Psychology, 56*(2), 239–245.

POWERS, J.R. (1996). *Rape investigation manual: A guide to investigative procedures, victim care, and case development.* Boulder, Colorado: Paladin Press.

RESICK, P.A. (1988). *Reactions of female and male victims of rape and robbery* (Final report, grant #85-IJ-CX-0042). Washington, DC: National Institute of Justice.

ROBERTS, A.R. (1997). Sexual assault and violence: Trends, policies, and programs. *Social work in juvenile and criminal justice settings.* Springfield, IL: Charles C. Thomas.

ROCHMAN, S. (1991). Silent victims. *The Advocate, 582,* 38–43.

ROTH, S., WAYLAND, K., & WOOLSEY, M. (1990). Victimization history and victim-assailant relationship as factors in recovery from sexual assault. *Journal of Traumatic Stress, 3*(1), 169–180.

ROTHBAUM, B.O., FOA, E.B., RIGGS, D., MURDOCK, T., & WALSH, W. (1992). A prospective examination of post-traumatic stress disorder in rape victims. *Journal of Traumatic Stress, 5*(3), 455–475.

RUSSELL, D.E. (1990). *Rape in marriage* (3rd ed.). New York: MacMillan.

SCARCE, M. (1997). *Male on male rape: The hidden toll of stigma and shame.* New York: Insight Books.

SIMON, L.J. (1996). Legal treatment of the victim-offender relationship in crimes of violence. *Journal of Interpersonal Violence, 11*(1), 94–106.

SOLOMON, Z. (1989). A three year prospective study of post traumatic stress disorder in Israeli combat veterans. *Journal of Traumatic Stress, 2*(1), 59–73.

ULLMAN, S.E. (1996). Correlates and consequences of adult sexual assault disclosure. *Journal of Interpersonal Violence, 11*(4), 554–571.

U. S. DEPARTMENT OF JUSTICE. (1996). *Bureau of Justice Statistics, National Crime Victimization Survey.* Washington, DC: U.S. Department of Justice.

U. S. DEPARTMENT OF JUSTICE. (1997). *Bureau of Justice Statistics, National Crime Victimization Survey.* Washington, DC: U.S. Department of Justice.

VAN DER KOLK, B. (1987). *Psychological trauma.* Washington, DC: American Psychiatric Press.

VAN DER PLOEG, H., & KLEIJN, W. (1989). Being held hostage in the Netherlands: A study of long term after affects. *Journal of Traumatic Stress, 2*(2), 153–169.

WAIGANDT, A., & PHELPS, L. (1990). The effects of homicides and suicides on the population longevity of the United States. *Journal of Traumatic Stress, 3*(2), 297–304.

WILSON, J. (Ed.). (1989). *Trauma transformation and healing: An integrative approach to theory, research, and post traumatic therapy.* New York: Brunner/Mazel.

WYATT, G.E. (1992). The sociocultural context of African American and White American women's rape. *Journal of Social Issues, 48*(1), 77–91.

ZILBERG, N., WEISS, D., & HOROWITZ, M. (1982). Impact of event scale: A cross-validation study and some empirical evidence supporting a conceptual model of stress response syndromes. *Journal of Consulting and Clinical Psychology, 50*(3), 407–414.

5

Police Involvement in Child Maltreatment

Multidisciplinary Child Abuse Investigations

Janet R. Hutchinson

❖

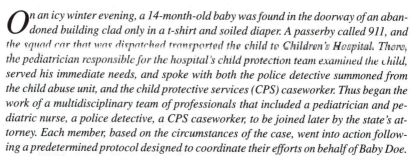

*O*n an icy winter evening, a 14-month-old baby was found in the doorway of an abandoned building clad only in a t-shirt and soiled diaper. A passerby called 911, and the squad car that was dispatched transported the child to Children's Hospital. There, the pediatrician responsible for the hospital's child protection team examined the child, served his immediate needs, and spoke with both the police detective summoned from the child abuse unit, and the child protective services (CPS) caseworker. Thus began the work of a multidisciplinary team of professionals that included a pediatrician and pediatric nurse, a police detective, a CPS caseworker, to be joined later by the state's attorney. Each member, based on the circumstances of the case, went into action following a predetermined protocol designed to coordinate their efforts on behalf of Baby Doe.

Within 48 hours, the toddler was treated and released to a temporary, emergency foster family under the supervision of the CPS caseworker. The police detective had located the child's family and begun the investigation that would lead to an initial charge of child endangerment against the baby's caretaker-father. By the third day, the baby and his three elder siblings had been placed with a nearby relative while the investigation continued, and their mother, a prostitute, was located. Within a week, a case conference was conducted with team members, including the prosecutor. After reviewing the information collected by the detective and CPS worker, which included the recommendations of the pediatrician who first examined the child, and school officials with knowledge of the elder siblings, the prosecutor concluded that there was sufficient evidence of both child abuse and neglect to bring charges against the caretaker-father. Ultimately, the *children were placed in long-term permanent foster care with extended family members, and the father was given a 12-month jail sentence on child endangerment charges. The outcome in this abbreviated scenario was a generally satisfactory one, but might have been quite different, indeed, tragic, had the collaboration among key child protection professionals not occurred.*

INTRODUCTION

This chapter discusses multidisciplinary child abuse investigation teams and the importance of police/law enforcement agency participation in the investigation and prosecution of child maltreatment cases. Several issues are discussed that confound the development and maintenance of multidisciplinary teams, not the least of which is cross-disciplinary cooperation. Cooperation is difficult under the most congenial circumstances, and cooperative ventures across disciplines present significant challenges. These challenges are compounded by the sense of immediacy that is present when investigations involve crimes against children, and when the professionals involved have conflicting objectives. Nevertheless, the importance of multidisciplinary alliances has been repeatedly demonstrated in communities across the United States, and the several models that have emerged for facilitating these alliances will be reviewed.

Child maltreatment, the preferred term for child abuse, encompasses four broad categories that are specific to harmful acts, or the failure to act, on the part of parents or caretakers. These categories are emotional abuse, neglect, physical abuse, and sexual abuse. The victim in these circumstances is a child, and since the victim is often an infant or very young child, he or she may be unable to participate as an older child or adult might in the investigation and prosecution of the alleged perpetrator. Because of their particular vulnerability, expert professionals long involved in the investigation and treatment of these young victims have prescribed a multidisciplinary, community-based approach as the best method for ensuring that the facts of the case are identified, that the child is safe, that due process is exercised, and that treatment is assured.

This approach, which will be discussed at greater length later in this chapter, is also designed to lessen the trauma experienced by young victims at the hands of the professionals who are expected to help them. Abuse may continue unabated when professionals fail to identify life-threatening acts against the victim because they are not trained to recognize known indicators, or when professionals, unskilled in investigative procedures, fail to obtain substantiating evidence that could lead to removal and prosecution of the abuser. Multiple interviews conducted by people who are not skilled in forensic interviewing of children also contribute to continued victimization. Clumsy, contradictory, and patronizing questions, and particularly leading questions, confuse the victim and trivialize the child's actual experience, possibly causing the invalidation of a child-victim's testimony. Saywitz, Nathanson, and Snyder (1993) emphasize that the questioner's communicative competence is of paramount importance in establishing the child's credibility as a reliable reporter. "Insufficient developmental sensitivity by professionals (as a result of lack of training or the adversarial role) can frustrate children trying to answer questions they are not yet capable of understanding" (p. 60). Deleterious outcomes caused by responsible professionals, because they fail to use expertise which can be readily gained through multiple means, can be described as iatrogenic effects, a term used to describe doctor-induced harm to a patient.

Such effects, applied to professionals involved in child maltreatment, may be mitigated by interdisciplinary training and cooperation.[1]

Exacerbation of the effects of child maltreatment may also occur when appropriate treatment is withheld as a result of broader system failures. Such failures include contradictory policies among state and community-based agencies charged with the responsibility for investigating, prosecuting, and treating child maltreatment cases. They also include inadequate staffing for units charged with the investigation and treatment of abuse victims and their family members, as well as too few or nonexistent supportive and treatment resources such as therapeutic day care and foster families, addiction rehabilitation services, therapeutic counseling and psychiatric services, and job programs to relieve the stress of unemployment and homelessness. These problems are difficult to remedy by individuals acting alone, or through the efforts of a single agency; however, cooperative interdisciplinary teams provide a ready forum in which skill development and resource issues may be discussed and perhaps solved.

RECOGNITION OF CHILD ABUSE AND MULTIDISCIPLINARY TEAMS IN LAW

Child maltreatment was formally recognized in the United States in 1874, when public acknowledgment of the abuse of a child named Mary Ellen led to the formation of the first Society for the Prevention of Cruelty to Children. But it was not until 1961, when physician C. Henry Kempe coined the diagnostic term *battered child syndrome,* that child abuse became recognized as a condition for which public policies were required. Kempe and his associate, Ray E. Helfer, published several books describing their research on battered child syndrome and developed the first newborn research project in the new field of child abuse prevention during the mid-1970s. Kempe and his associates also created the first multidisciplinary team based in a Denver, Colorado, hospital (1959) in recognition of the need for coordination among those professionally involved in the investigation and treatment of abused children. By 1967, every state had enacted laws requiring that certain professionals report suspected child abuse, thereby greatly increasing the number of reports of child maltreatment, annually. In 1974, the Child Abuse Prevention and Treatment Act (CAPTA, Public Law 104-235, amended and reauthorized in 1996) enacted by the U.S. Congress required that states receiving federal funds under its provisions also establish multidisciplinary teams. Finally, by 1986, the Children's Justice Act provided funds to states to establish child abuse task forces that would include advocates, representatives of child protective services, health and mental health services, police/law enforcement, legal services, the court, and client/parent participants to review and evaluate methods of handling child abuse cases.

Public departments of social services are required to investigate reports of child maltreatment where the alleged perpetrator is a family member or caretaker; however, abuse that occurs outside the family and that is not connected with the family may be investigated by police/law enforcement personnel. Police routinely intervene in domestic violence

---

[1]For information on forensic interviewing, contact the National Children's Advocacy Center, 200 Westside Square, Suite 700, Huntsville, AL 35801, or the National Clearinghouse on Child Abuse and Neglect, 330 C Street, SW, Washington, DC.

cases, reporting evidence of child maltreatment to child welfare services. However, studies indicate that from 40% to 60% of domestic violence cases also involve the abuse of a child (Dykstra & Alsop, 1996; English, 1998; National Clearinghouse on Child Abuse and Neglect, n.d.). Straus and Gelles (1990) report that battered women may also batter their children, and Anderson (1999) suggests that for children, witnessing repeated acts of domestic violence is a form of psychological abuse. Given these findings, the role of policing/law enforcement is, by implication, expanded to include interviewing children in domestic violence cases to determine whether they too are victims. Both departments of social services and police/law enforcement agencies differ in structure—the degree of specialization may be determined by the size of the agencies and the communities they serve. Generally, child maltreatment investigations are conducted by CPS units in departments of social services. The exact circumstances that define the investigative roles of departments of social services and police/law enforcement are statutorily defined by the states.

THE EXTENT OF CHILD MALTREATMENT IN THE UNITED STATES

The U.S. Department of Health and Human Services compiles and reports state statistics on child maltreatment.[2] According to reports for 1997, 2,943,829 children were the subjects of referrals for investigation in the 50 states and the District of Columbia. Forty-three states reported that 825,131 cases of abuse and neglect were substantiated. Also in 1997, the states (n=39) reported that the largest class of perpetrators was parents (401,598), followed by other relatives (54,573, n=38), foster parents (2,733, n=37), facility staff (1,714, n=31), and child care providers (5,199, n=35), noncaretakers (25,621, n=35), and unknown perpetrators (41,294, n=32) constituting the remaining groups. State victim data (n=51) indicates that 56% of investigations involved neglect; 25% involved physical abuse; 13% involved sexual abuse; and 19% were victims of emotional maltreatment, medical neglect, or some other form of maltreatment. As of March 31, 1998, 520,000 children were in foster care, the highest number recorded since the early 1970s. These statistics give some indication of the magnitude of the problem facing local and state agencies charged with responsibility for investigating and treating victims of child maltreatment.

MODELS OF MULTIDISCIPLINARY TEAMWORK

The American Public Welfare Association[3] and the Police Foundation (Sheppard & Zangrillo, 1996) together conducted a study of randomly selected child protective services and police/law enforcement agencies during 1991 and 1992. The study was designed to assess the methods and models of joint investigation used by child protective and police/law enforcement agencies. The researchers identified three general models used in the coordina-

---

[2]Totals exceed 100% due to multiple incidences. Not every state reports in every category. The class of perpetrators is indicated by the first number in the parenthesis; the n is the number of states in that category. Data may be retrieved from the US/DHHS Home Page: acf.dhhs.gov/programs/cb/ncanprob.htm

[3]The American Public Welfare Association has changed its name to the American Public Human Services Association.

tion of child abuse investigations. In the first model described by the authors, joint investigations are conducted by specialized investigative units of police or sheriff's departments, along with CPS workers, and designated district attorneys using written protocols and community facilities to interview alleged victims. A second model uses centers where multidisciplinary team members including police detectives, CPS caseworkers, and others, depending upon the circumstances, meet to conduct victim interviews and develop strategies for conducting the case. These centers, according to the report, have limited staffing and part-time administrators. The third model described by the authors is the child advocacy center which operates an independent, full-service multidisciplinary program staffed by victim advocates, specialists in forensic interviews, mental health therapists, and administrative support staff. A number of variations on these models were identified among the 325 municipal police agency, 279 county police/law enforcement agency, and 239 child welfare agency survey respondents.[4] Whether one of these models was used, or alliances between agencies were informally constituted, most respondents indicated that investigations are conducted jointly between child protective services and police/law enforcement agencies. The common objective shared by the models described is the joint interview, the purpose of which is to minimize the number of interviews to which child victims are subjected.

ISSUES IN THE DEVELOPMENT AND MAINTENANCE OF MULTIDISCIPLINARY TEAMS

The elements have long been in place for establishing and maintaining a coordinated presence in local investigations of child maltreatment complaints; however, multidisciplinary coordination of the agencies and organizations that play central roles in the investigation and treatment of child maltreatment cases is by no means universal. The Office of Juvenile Justice and Delinquency Prevention (OJJDP) has produced several helpful pamphlets for police/law enforcement professionals on the subject of child maltreatment. One of them, *A Portable Guide to Investigating Child Abuse* (1997), begins with the statement that "Child abuse is a community problem. No single agency has the training, manpower, resources, or legal mandate to intervene effectively in child abuse cases. No one agency has the sole responsibility for dealing with abused children." However, it could be argued that child maltreatment is *not* perceived as a community problem (at least, not by members of the community at large), nor has the agency that is charged with the initial response to intrafamilial child maltreatment complaints, the Department of Social Services, necessarily viewed other agencies as having other than an ancillary role in the investigation and treatment process. For example, if child maltreatment is truly viewed as a community problem, child welfare agencies would be adequately staffed with trained professionals. Caseloads would be small enough to permit adequate planning for each child and family assigned to a caseworker. Similarly, if child maltreatment was considered a community problem, police/law enforcement agencies would have adequate numbers of trained personnel assigned to dedicated child abuse investigation units. Police academies would routinely train new officers to recognize and respond appropriately to cases of child maltreatment, and forensic interviewing with children would be a

---

[4]The article reporting on this study does not give the total number of persons or organizations surveyed. Attempts to identify this number through the sponsoring organizations were unsuccessful.

staple on the training menu in preparing police/law enforcement officers for their jobs. Furthermore, if child maltreatment was truly considered a community problem, state's attorneys would not only be willing to participate in multidisciplinary teams, they would uniformly accept a leadership role in ensuring that substantiated child maltreatment cases are prosecuted.

In virtually every community in the United States, the public child welfare agency is charged with investigating complaints of child abuse and neglect (child neglect constitutes the greater number of complaints lodged with public child welfare agencies). Without a strong commitment from child welfare to include other agencies in the process, multidisciplinary teamwork has little chance of success. How and when the actors in the child maltreatment drama share responsibility for the welfare and fate of abused children and their families is in great part dependent on the commitment to action by the executives of child welfare agencies and their counterparts in police/law enforcement, the state's attorney and the courts, and health and mental health services. It is with the leadership that much of the difficulty in establishing viable teaming approaches lies. A recent survey[5] of Virginia county agencies suggests that where teams are working well, there is a strong leadership presence, most often by the state's attorney's office. Leadership from this quarter is appropriate since it is through the state's attorney that cases are prosecuted. Furthermore, close involvement by the state's attorney can assure that others on the multidisciplinary team collect evidence in a manner that facilitates prosecution when warranted.

The failure of teamwork seems most frequently to result from the loss of a key leader and the disinterest of his or her replacement, often an elected official or political appointee. It is striking how significant to the process one key member can be. In areas where multidisciplinary teams have not taken root, it has been attributed to the absence of committed leadership resulting in an unwillingness or inability to participate by a lead agency. When this agency is the Department of Social Services, access to the victim is cut off from other agencies in the community, often behind a shield of confidentiality, unless the abuse results in a child's death. When the noncontributing agency is police/law enforcement, the CPS worker is left without the legal authority to remove the perpetrator from the child's home, leaving removal of the child as the only recourse. When the state's attorney is a nonparticipating agent, the police can do little to influence prosecution, and child welfare once again lacks other recourse beyond removal of the child. Furthermore, although placing a child in foster care may seem the most reasonable action by other agencies and the community, foster care is not a guarantee of safety, nor is it always the best choice for ensuring a child's long-term stability. Children who have experienced physical and psychological maltreatment often experience multiple foster home placements, resulting from foster parents' inability and unwillingness to work with and endure the behavioral and emotional challenges that such children bring.

As important as leadership is to the success of multidisciplinary teams, other factors also impede multidisciplinary work. For example, interdisciplinary cultural differences contribute to miscommunication and misunderstandings among team members. Language (jargon) and customs (established behaviors) define the membership and boundaries of a discipline (branch of learning as in criminal justice, law, social work); these constitute the sociology of the discipline. Disciplinary boundaries create barriers among potential team

---

[5]Discussed in an interview with Dorothy Hollohan, Virginia Department of Criminal Justice Services, October 1999.

members when someone invokes perceived privileges and status in order to elevate their own importance over others. Contradictory value systems are another impediment to teamwork, particularly when team members confront problems over which they fundamentally disagree on the course of action. A common point of disagreement is whether or not to remove the child from his or her home. The removal of a child may seem the most expeditious course of action to a police/law enforcement officer or emergency room physician; however, the social worker may believe that the best interests of the child are served by providing in-home support to a nonabusive caretaker. A persistent unwillingness to accommodate perspectival differences leads to the marginalization of one or more participants who may not be viewed as having equal standing among the other professionals typically involved in child maltreatment cases. As a consequence, the breakdown that occurs deprives child maltreatment cases of professional interventions that might otherwise benefit both the child victim and his or her family members.

Also problematic are the resource-related issues that restrict a team member's availability when he or she is needed on a case. Selective availability due to staff shortages may be the result of high staff turnover or, when turnover is not an issue, because the agency is simply understaffed or, in police/law enforcement agencies, because child abuse investigations are not a priority. Staff shortages, attributable to underfunding, hiring freezes, and low pay, are endemic to child welfare and police/law enforcement agencies. Unfortunately, the case that is not acted upon today may well become tomorrow's crisis—or worse, a statistic. Related to these problems is inadequate training of child abuse investigators, both caseworkers and police officers. Agency administrators faced with severe budget limitations are inclined to eliminate training or staff development rather than client services, particularly when the investment in training is lost through high worker turnover. Workers with excessive caseload demands may forgo training offered outside the agency rather than fall further behind in their work. When one or more members are not trained to do their own jobs and are unclear about the roles of other team members, teamwork becomes burdensome for more experienced personnel. Under these circumstances, potential team members may find that it is easier to do the job alone. This, however, is an inadequate solution, since acting alone deprives the child-victim and his or her family the full range of supportive services that would otherwise be available to them. High worker turnover and lack of preparation for the job through training is generally, at root, a budget problem which, in most circumstances, requires a political decision for resolution. When crime becomes a community concern, more officers are hired; social service agencies have not experienced this kind of political support since the Great Society of Lyndon Johnson's presidency emphasized antipoverty programs.

Two common reasons for high turnover were noted previously; however, turnover is also related to the characteristics of the job itself. Child abuse investigations are commonly ranked with domestic violence calls on the scale of preferred duties among many police/law enforcement officers; that is, they are not the duty choice. Cases are often difficult to substantiate, difficult to prosecute, personally painful, and emotionally draining. In some jurisdictions, it is the first assignment for rookies, and the last choice among detectives. One veteran detective interviewed for this chapter indicated that several positions in his small child crimes unit have been vacant for some time; apparently few people are interested in the job. Nevertheless, given the sensitivity of the task and the level of commitment required to sustain a healthy emotional state while investigating child maltreatment, it is important that officers and caseworkers alike not be forced to do investigations against their will. Lanning and Hazelwood (1988, p. 121) caution that officers assigned to investigate sexual crimes should be volunteers who are carefully screened and trained, and that the assignment should never be used as discipline or punishment. Marginal performers should not be expected to deal with highly intrusive, violent crimes where their initial interactions can have a direct influence on the victim's feelings of self-worth. This excellent advice applies to other crimes of maltreatment against children as well.

Police officers who investigate child abuse frequently face difficulties with their peers, jokes and name-calling, innuendo about their sexual preferences, isolation from and the derision of their peer officers, as well as the stress of being continuously exposed to child-victim crimes, particularly cases of child sexual molestation (Lanning & Hazelwood, 1988). CPS workers share the stress that results from repeated exposure to child abuse and neglect; however, peer issues are more likely to occur outside the agency when their status as experts is challenged. Caseworker stress is also more likely to result from excessive caseloads, contradictory expectations in their agencies and the community, and poor preparation for their jobs.

INTERDISCIPLINARY COOPERATION AND THE ART OF HESITATION[6]

Cooperation is the operative term in building a multidisciplinary investigation and treatment program. To cooperate suggests a give-and-take process that many professionals working across disciplines find challenging. It requires a degree of hesitation in working with others—hesitating long enough to hear and interpret meanings in the words and actions of the other. Each member of a team brings knowledge and expertise that contributes to the preinvestigation planning, investigation, and treatment phases of a case. Each has a set of objectives that he or she is concerned with achieving. For example, everyone on the team is concerned for the victim's safety; however, the police/law enforcement officer's and state's attorney's objectives are short-term and very specific, whereas the objectives of child welfare have a longer timeline.

For example, when considering the circumstances of the child's caretaker, his or her capacity to protect the child is of critical importance. If the perpetrator is ordered to leave, is the remaining caretaker capable of enforcing the order? Is the caretaker using alcohol or drugs? In considering the perpetrator: Is he or she a family member or someone living on the premises? Is the alleged perpetrator using alcohol or drugs? Does he or she have a history of violence? In considering the victim: Does the child need medical attention? How old is the child? Is the child able to describe the circumstances of the abuse? Is the child physically or mentally disabled? Is the child safe at home or must the child be removed from the premises? If the child must be removed, is there a safe relative placement immediately available or will it be necessary to locate an emergency foster care placement? If an emergency placement is available, about how long will the child need the placement until he or she can return home, or to the home of a relative? In considering the abuse: Does it appear that the complaint is credible? Is this a repeated abuse, or does it appear that this is the first

---

[6]Credit is given to David J. Farmer for his application of the art of hesitating to administrative situations, which he discusses in several of his recent publications, the most recent of which is referenced.

incident? If it appears to be the first incident, are the family circumstances such that it is likely to occur again? In determining whether or not to request police assistance, the CPS investigator must judge the likelihood that in this case a criminal law violation has occurred. Statute generally stipulates which types of cases are to be reported to the state's attorney and local police/law enforcement agency. Abuse or neglect involving the death of a child is such a case. Other instances may include injury or threats of injury to a child during the commission of a felony or class 1 misdemeanor, sexual abuse, suspected sexual abuse or other sexual offenses involving pornography, abduction of a child, drug offenses involving children, and contributing to the delinquency of a minor.

It should be noted that CPS investigators do not generally work alone. They routinely consult with supervisors and coworkers; however, the press of large caseloads and the demands of other cases on limited resources may determine a course of action over which other professionals in the community may disagree. Furthermore, the current practice wisdom expects a great deal of the child welfare system. It is expected to protect children and keep them safe from harm while preserving the family for the child whenever possible. The language of the CPS worker is multiple. For other members of the multidisciplinary team to recognize the contextual circumstances under which fellow members operate makes it possible to operationalize hesitancy.

Police and law enforcement officers generally view their role in the investigation of child abuse complaints as limited in scope. Police/law enforcement officers are expected to be knowledgeable about rules of evidence—their principal task is to establish whether a crime has been committed, and if so, to remove the perpetrator, collect evidence of the crime, and refer the case to the state's attorney. In the words of one detective interviewed, "My role is to cut the perp[etrator] out of the picture and let social services provide services to the rest of the family, like counseling, money." The county in which he works does not have a multidisciplinary team, although the principals work well together. An attempt to develop a child advocacy center failed when the county council objected to the cost. Among the frustrations that he has experienced with social services is the high turnover among CPS investigators and what he describes as the inadequate training that new investigative workers receive before going out on the job. Certainly, an advantage of the multidisciplinary team is the knowledge that passes from one member to the other, and the learning process that occurs as a result. This has been described as organic collaboration. In their collective discussion and criticism, each member brings to the table his or her particular expertise—raising the level of expertise of each to the most skilled of its members. Gramsci (1992, pp. 28–29) writes that " . . . in this kind of collaborative activity, each task produces new capacities and possibilities of work, since it creates ever more organic conditions of work. . . ."

The factors cited here would seem to support one of the models described earlier in the study conducted by Sheppard and Zangrillo (1996). Colocating key investigation and treatment personnel in a facility designed to accommodate the special needs and circumstances of child-victims has the potential to eliminate several of the problems cited as factors impeding interdisciplinary investigations. Those who staff such centers ideally work together to overcome the differences inherent in their respective disciplines; they learn through close association with one another to interpret and adapt to each others' language and customs as Gramsci proposed. They are there by choice. Furthermore, a commitment of resources by a community to an advocacy center may mitigate those problems related to dependence on the leadership of a single individual to maintain a viable but informal multidisciplinary team relationship.

There are several potential disadvantages of the advocacy center approach. One is the potential for overdetermining an investigation, that is, setting in motion a process that leads to substantiating abuse, particularly sexual abuse, with insufficient evidence. Although sexual abuse, most will agree, is a particularly heinous crime, it is also difficult to substantiate. The advocacy center approach was prompted by the belief that very specialized methods for interviewing are necessary to ensure that the child-victim's testimony is obtained by skilled professionals. Optimally, advocacy centers are staffed by trained specialists. When they are not, overdetermination is a distinct possibility. Another disadvantage of the advocacy center approach is the potential for elevating one type of maltreatment, sexual abuse (an estimated 13% of cases), over other types of physical abuse, as well as physical and emotional neglect, and the diversion of scarce funds away from these equally important and serious types of maltreatment. Also, unless colocated with an emergency medical facility, such centers do not obviate the need for medical treatment and the collection of physical evidence by trained medical personnel. Furthermore, many jurisdictions are simply unable to summon the financial resources to build, staff, and maintain separate facilities for the investigation and treatment of child victims. In geographically large jurisdictions, locating a facility for optimal access is a challenge. It seems reasonable that the particular approach used by a community of service providers be tailored to community circumstances and that these circumstances are best understood when a commitment is made to work cooperatively in facilitating the investigation and treatment processes.

CONCLUSION

In this chapter, the legal and professional bases for developing multidisciplinary child protective service teams have been discussed as have several models that are in use in communities around the country. The circumstances that impede their successful development and maintenance are both specific to the disciplines involved and can be viewed as the sociology of the discipline. There are also structural impediments, some of which may be resolved through simple adjustments in agency policies, others of which require political remedies. It is true that many child maltreatment cases are dealt with by child welfare agencies without the involvement of police/law enforcement or the state's attorney. However, for the relatively small percentage of cases in which criminal prosecution is warranted, both police/law enforcement agencies and the state's attorney are expected to be willing and knowledgeable participants. When either one or both do not do their jobs, the consequences may indeed be fatal. Both are influential forces in the prosecution of child abusers and are potentially influential leaders in the community of professionals. It seems particularly the case that in communities where the state's attorney is not an active participant, multidisciplinary teams are less likely to succeed.

Several models of multidisciplinary cooperation have been discussed in this chapter; however, no particular model is necessarily the best for a given community. In some communities, medical professionals have been the driving force in developing alliances among professionals involved in child maltreatment; in others, the state's attorney has taken the lead. In others, child advocacy organizations have raised community awareness to the level of action on behalf of abused children. In still others, victim advocates have been the organizing force. In jurisdictions unable to support a child advocacy center or similar model, efforts to formalize a multidisciplinary process are preferable to the more common informal alliances since the latter are subject to the instability that results when a significant actor in the alliance is lost to the process. Recognizing the predictable objections to interjurisdictional approaches, they are, nonetheless, a proposition worth investigating. Precedent exists with regional medical and mental health facilities, state police, and regional medical examiners. Abuse does not respect county lines, and its victims should not fall hostage to jurisdictional claims. Pooling resources wherever and whenever possible simply makes good sense.

Although very little outcome research has been conducted on the utility of the models discussed, anecdotal information and years of experience by professionals working with abused children and their families suggests that strong alliances reduce the trauma that children, and nonoffending caretakers, experience as a result of professional intervention. Formal alliances may also benefit professionals. They can become a forum for discussing and perhaps resolving issues common to one or more service agencies in the community, and they can become a source of knowledge transfer among more experienced and less experienced participants—organic collaboration. In the last two decades, studies conducted with large samples have given us sufficient data to support claims that child maltreatment is a very serious social problem that transcends race, class, education, and income. A brief tour of the Internet reveals that many excellent resources currently exist to help individuals, professionals, agencies, and communities deal with what appears to be growing incidences of serious violence against children. Despite the ease with which these resources can be found, and our sophisticated knowledge of the physical, psychological, and social harm to children who experience violence, there are still too many communities for which multidisciplinary investigation and treatment of victims of child maltreatment is not a policy priority.

REFERENCES

ANDERSON, C.L. (1999, December). *The co-occurrence of child abuse and domestic violence.* Presentation at the Improving Investigation and Prosecution of Child Abuse Conference, Virginia Beach, VA.

NATIONAL CLEARINGHOUSE ON CHILD ABUSE AND NEGLECT. (n.d.). *Double jeopardy: Domestic violence and child maltreatment.* Washington, DC: U.S. Department of Health and Human Services.

DYKSTRA, C.H., & ALSOP, R.J. (1996, Spring). *Domestic violence and child abuse* [Monograph]. Denver, CO: American Human Association.

ENGLISH, D. (1998, March). *Co-occurrence: Child abuse and domestic violence.* Presentation at the Sixth Forum on Federally Funded Research on Child Abuse and Neglect. Bethesda, MD.

FARMER, D.J. (1999). The discourse movement: A centrist view of the sea change. *International Review of Public Administration, 4*(1), 3–10.

GRAMSCI, A. (1992). Prison notebooks. In Q. Hoare & G.N. Smith (Eds. and Trans.), *Selections from the prison notebooks.* New York: International Publishers. (Original work written between 1929 and 1935)

KEMPE CHILDREN'S CENTER. *History of the Kempe Center* [On-line]. Available: http://www.kempecenter.org/history.htm

LANNING, K.V., & HAZELWOOD, R.R. (1988). The maligned investigator of criminal sexuality. *FBI Law Enforcement Bulletin, 57*(9), 1–10.

OFFICE OF JUVENILE JUSTICE AND DELINQUENCY PREVENTION. (1997). *Law enforcement response to child abuse: A portable guide to investigating child abuse* (NCJ Publication 162425). Washington, DC: U.S. Department of Justice.

SAYWITZ, K.J., NATHANSON, R., & SNYDER, L.S. (1993). Credibility of child witnesses: The role of communicative competence. *Topics in Language Disorders, 13*(4), 59–78.

SHEPPARD, D.L., & ZANGRILLO, P.A. (1996, Winter). Coordinating investigations of child abuse. *Public Welfare, 21*–32.

STRAUS, M.A., & GELLES, R.J. (1990). *Physical violence in American families: Risk factors and adaptations to violence in 8,145 families.* New Brunswick, NJ: Transaction Publishers.

INTERNET LINKS

Office of Juvenile Justice and Delinquency Prevention, National Network of Children's Advocacy Centers, Inc.
http://www.ojjdp.ncjrs.org/pubs/trngcatalg/nncac-pp.html

American Bar Association Center on Children and the Law
http://www.abanet.org/child/home.html

National Clearinghouse on Child Abuse and Neglect Information
http://www.calib.com/nccanch/

American Professional Society on the Abuse of Children
http://www.apsac.org/

The Child Abuse Prevention Network
http://child.cornell.edu/

National Children's Advocacy Center
http://www.ncac-hsv.org/

National Association of Counsel for Children
http://naccchildlaw.org/

National CASA Association
http://www.nationalcasa.org/

6

Victim Interviewing in Cases of Domestic Violence

Techniques for Police

Denise Kindschi Gosselin

INTRODUCTION

The literature is filled with debates on every aspect of domestic violence from definitions to prevalence rates. Academics, feminists, and policy makers disagree, and agree to disagree, on research methodologies and control strategies. However, the controversy does not actively involve police officers whose actions are governed by legislation and police department policy. Regardless of the prevailing winds, police officers are expected to provide the first response to cases involving domestic abuse. Police officers must respond decisively and without emotion, ready to handle any level of violence that they encounter. At the same time the officers are expected to act with sensitivity and impartiality. The question must be asked, are they being adequately prepared to handle the challenge? Training academies do not focus on domestic violence for as long as it truly deserves. Rarely is a separate interviewing course offered during recruit training. Even rarer is one that addresses interviewing for domestic violence cases. It is no wonder that police officers view these cases as troublesome; they have few resources on how to handle these difficult situations.

Arming our police officers with the ability to question those involved in the domestic violence case might be the best weapon that can be furnished. Interviewing is a basic skill that should accompany every police officer who responds to domestic abuse. The ability to talk to the people involved sets the stage for a favorable resolution. This chapter is written in hopes of making a contribution to the effort. For students of criminal justice it should supplement other courses on domestic violence. Some myths and misconceptions about police officer response and expectations may be shattered. My goal is to persuade persons involved in domestic violence response that interviewing options do exist and that they should be exercised. It is in the best interest of the victims and police officers to talk to each other. The dialogue can be stymied or encouraged by the officers at the scene.

WHAT IS DOMESTIC VIOLENCE?

There is little definitional agreement regarding domestic violence. The National Council of Juvenile and Family Court Judges' *Model Code on Domestic and Family Violence* defines domestic violence[1] as one or more of the following acts:

1. Attempting to cause or causing physical harm to another family member or household member,
2. Placing a family or household member in fear of physical harm, or
3. Causing a family or household member to engage involuntarily in sexual activity by force, threat of force, or duress.

For the purposes of this chapter, domestic violence includes the *Model Code* definition and is expanded to include persons in a dating or intimate relationship regardless of where they reside.

Spousal abuse is most frequently cited, but same-sex relationships and parent-child strife may be as prevalent. Every state and the District of Columbia have enacted laws that forbid domestic violence and provide civil remedies for the victim of abuse, but each defines the domestic relationship differently. Both federal and state law recognize that domestic relationships include partnerships where people live together, but are not legally married. Persons who are related by blood or through marriage are typically protected. Some states identify roommates and dating couples as constituting a domestic relationship. Domestic brutality is committed against both women and men and includes violence in gay and lesbian relationships. Most legislation is written in language that is gender-neutral, providing protection regardless of the gender relationship. It should be noted that the laws that allow a victim to apply for a restraining order specifically exclude same-sex relationships in seven states[2] (National Coalition of Anti-Violence Programs, 1998).

The term *domestic violence* describes the relationship of the perpetrator to the victim, not the specific crime that has been committed. Domestic abuses often constitute criminal code violations. Acts that are illegal if committed against a stranger are also prohibited against an intimate or family member. Assault, aggravated assault, rape, and murder are a few examples of crimes that may be domestic. Stalking is a crime recently recognized as being associated with intimate violence.

---

[1]*Model Code* 1-2, s 102 (1994).

[2]These states include Arizona, Delaware, Louisiana, Montana, New York, South Carolina, and Virginia.

HOW OFTEN DOES DOMESTIC VIOLENCE OCCUR?

It is difficult to determine with exactness the frequency of domestic abuse. Varying definitions of violence and age of victims make comparisons difficult. In addition, official statistics based on arrest rates and self-report studies are often conflicting. Research studies that suggest equal rates of victimization for men and women are often criticized for their failure to identify those women who acted in self-defense.

Apart from the controversy, a few generalities and trends have emerged. We know that victims of abuse cannot be identified through characteristics such as age, gender, or financial status. Neither the victims nor the victims' actions cause domestic violence. We cannot blame victimization on victims' alcohol or drug use. We cannot excuse batterers because they may drink excessively or abuse drugs. What may be a coping mechanism for the victim often is cited as an excuse for the perpetrator. Batterers are bullies who prey on others. Excessive needs for power and control have been identified as the core of their purposeful abusive actions.

On average each year from 1992 to 1996, approximately 8 in 1,000 women and 1 in 1,000 men age 12 or older experienced a violent victimization perpetrated by a current or former spouse, boyfriend, or girlfriend (Greenfeld et al., 1998). Annual rates range from 12 per 1,000 for marital rape to 116 per 1,000 for any adult domestic violence act (MacNeil et al., 1998).

The prevalence of husband battering is an issue of considerable controversy; its very existence is often denied (Hattendorf & Tollerund, 1997). The *Violence by Intimates* study found a prevalence rate for domestic abuse at approximately a 7 to 1 ratio; 1 million females versus 150,000 males were identified as victims of intimate violence (Greenfeld et al., 1998). The accepted belief that women are more likely to sustain injuries as a result of domestic violence is being contradicted by some studies that suggest men are equally injured by intimate violence (MacNeil et al., 1998).

Relatively few studies have examined the prevalence of battering among lesbian couples and gay men. Rates of violence for lesbians range from 25% to 48% of relationships, comparable to that of heterosexual intimates (Coleman, 1996). The authors of *Men Who Beat the Men Who Love Them* hypothesize that battered gay men may experience domestic violence at a rate higher than heterosexual couples (Island & Letellier, 1991). A recent review of the literature concerning lesbian, gay, bisexual, and transgender domestic violence victimization suggests a prevalence rate between 25% and 33% (National Coalition of Anti-Violence Programs, 1998).

POLICE INTERVENTION

Police respond to violence that involves a spouse or lover up to 8 million times per year (Sherman & Rogan, 1992). At one time it was considered the most dangerous police call; now it is generally accepted as the most frequent form of violence in the United States. Emotions are extremely high during a dispute; this makes the situation volatile and unpredictable. For the safety of the officers, it is best when at least two police officers are sent to respond. If the situation is believed to involve weapons, more officers may be necessary.

The domestic dispute symbolizes everything that runs counter to the expectations of a crime fighter. In a domestic dispute it is not unusual for the parties to be hostile to police interference. Although police officers anticipate cooperative victims who are appreciative of their efforts to intervene, the opposite may be true for the domestic call.

Clearly, police have the duty to respond to reports of domestic disputes. A protocol for responding to the home of an ongoing dispute should be determined prior to the event. A screeching police siren and flashing lights may be necessary for the cruiser to move through traffic, but is not needed to gain entry. Turn off unnecessary lights and approach the house as quietly as possible. This allows the officers to listen—one of the most important parts of a successful response to domestic violence. Domestic abuse calls are unpredictable because no two quarrels are alike. The officers should be prepared to stand for a second or two at the side of the door or window and attempt to listen to what is going on inside. In this way it might be learned if the victim is in imminent danger. The absence of any noise coming from the home or apartment may have as much significance as yelling and screaming. The actions of the police officers often influence and shape the events that will follow their entry. In most cases it is not necessary to break down the door and blaze through the home with weapons drawn. It is important to knock and announce the police presence. The events, and not the emotions of the officers, should determine the speed of the actual entry.

It is not unusual for one person to answer the door and tell the police officers to go away, that everything is fine. The officers' duty cannot end with a dismissal, however. Public officials can be held personally liable under the Federal Civil Rights Act (42 U.S.C. 1983) for injury or death caused by the negligent or wrongful act or omission while acting within the scope of their office or employment. Similar state civil rights acts dictate the extent of liability for the failure to act as the result of policy or custom. It is necessary to at least view the other party to make sure that he or she is not injured.

All people present or involved should be interviewed. If the police officers are not permitted to view or speak with the victim in a dispute they should consider whether *exigent circumstances* exist with sufficient evidence to enter the home without permission in order to search for the victim. Exigent circumstances are those that demand unusual or immediate action. A judicially determined exception to the requirement that police officers obtain a warrant prior to entering a home is the occurrence of exigency—a "now or never" emergency. Protecting a victim of domestic violence from further abuse has been recognized as a dire emergency justifying police entry.[3] Here are a few factors to consider when making the determination of whether exigent circumstances exist: the presence of weapons; prior history of domestic violence or prior police calls to the residence; threats to kill; drug and/or alcohol use prior to the incident; existence of a restraining or no-contact order; the commission of a domestic crime; and/or suppression of a breach of the peace.[4] At one time, repeat calls to the same residence were considered a nuisance to police officers; now they should be recognized as a red flag for the possibility of domestic homicide (Johnson, Li, & Websdale, 1998).

Still there is no consensus on how police should respond to the dispute call. Should they make an arrest or not? Frequently the extent of harm that has been committed is the standard

---

[3]Refer to *People v. Higgins,* 26 Cal.App4th 247, 31 Cal.Rptr.2d 516 (Cal. App. 1994); *Commonwealth v. Rexach,* 20 Mass. App. Ct. 919, 920 (1985); *United States v. Bartelho,* 71F.3d436,441-442 (1st Cir. 1998).

[4]*United States v. Rohrig,* 98 F. 3rd 156, 1520-1521, 1996.

used to determine if an arrest should be made. The preference of the victim is often a determining factor. Some states require that police officers arrest a perpetrator in all domestic abuse allegations where probable cause exists to believe that a crime has been committed. Federal and state legislation govern the situations when an arrest must be made; the policy of the police department will guide police officers in the situations that are less clear.

Urging us to avoid the temptation of determining who is the most deserving, Philip Cook looks at the scant research surrounding male victimization and the dynamics of injury toward men (Cook, 1997). Female assailants against male partners show an increased use of deadly weapons, he reports, and of knives in particular. Though Cook does not minimize the violence against women, he also advocates the aggressive use of mandatory police arrest powers when men are the victims of physical assault. A recent charge is that the police and the courts do not take assaults by women seriously. In one prominent study to document domestic violence intervention from the victim's perspective, evidence was uncovered suggesting differential treatment by police toward male victims (Buzawa & Austin, 1993). None of the men who had been victimized were satisfied by the police handling of their assault, yet most of the female victims indicated that they were.

Once a safe and successful entry has been gained, it is time to investigate the situation. Preconceived notions and expectations should be thrown aside in order to conduct a nonbiased investigation. The challenge for the criminal justice community is in assigning blame. One party must be responsible and held accountable in the criminal justice paradigm. Some police officers who respond arrest both the parties in the domestic violence dispute to avoid having to sort it out. This practice of mutual arrest is strongly discouraged by the courts. Not only does it confuse the court on how to proceed, it often results in neither party being prosecuted. Victims are not being protected and batterers are empowered. In addition, this tactic serves to punish the victims of domestic violence because they are being put before the court as if they were the abusers. *Effective interviewing is the best way to gather the facts in the case.*

PURPOSEFUL INTERVIEWING

An estimated 80% of an officer's investigative work consists of obtaining information through interviews. It is common knowledge that the vast majority of cases are solved when someone talks about what happened. Good communication skills and a willingness to listen to people determine how successful the officer will be in obtaining a statement. An interview may be formal or informal; either way, the interviewer must retain control of the situation at all times.

Purposeful interviewing refers to the realization that statements are sought for specific reasons. Domestic violence crimes are complex, necessitating different approaches in order to elicit truthful information. There are often strong emotional ties between the victim and the offender; the victim may be financially dependent on the assailant. In some cases the perpetrator faces potential loss of employment if involved in domestic abuse. A victim may be unwilling or unable to provide information that will be used against the partner; hence purposeful interviewing aims to verify and give credibility to the victim without expecting a high level of cooperation. The reasons for purposeful interviewing can be classified into four general justifications: (1) victim identification, (2) risk assessment, (3) evidence gathering, and (4) outcome determination.

Victim Identification

Police officers presume readily identifiable victims and offenders; that is not the case in the domestic abuse dispute. Victim identification is the first hurdle that police officers must overcome through effective interviewing. It is relatively easy to detect the victim in almost every other form of violence that police officers encounter. When police officers harbor preconceived notions of how a victim of crime should act, the difficulty of making a true identification is compounded. Victims want to be treated with respect. The level of their cooperation can be tied directly to the treatment that they receive from officers who respond to the scene. One risks offending the victim without meaning to if the identification is based on a quick first impression. Taking the time to properly identify the victim necessitates treating all parties with respect.

One false assumption is that victims align themselves with the police officers who have "come to their rescue." In the domestic dispute, the victim may side with the perpetrator against the police. This can be confusing to responding officers. It helps to remember that there is a "relationship" between the victim and the abuser; the police are often the outsiders. The victim may be in love with the perpetrator, although she or he doesn't want to be battered. Issues of victimization within domestic relationships are complex, and it is recommended that police officers be trained in the dynamics of abuse. Even so, it is not up to the officers who respond to the dispute to make judgments on the relationship. Making assumptions without evidence is inappropriate within the context of police action. The following are some common examples of false presumptions in domestic violence situations:

False Assumption: The Injured Person Is the Victim.   In some cases the injured party may be the perpetrator. This happens when the victim uses force or weapons to defend her- or himself. A self-defending person will likely be extremely remorseful. After the individual has been disarmed the police officer should ask, "Why did you do that?" or "Didn't you realize that someone could be hurt?" A self-defending person will offer justifications such as: "He was going to kill me," "I had no choice, she came at me with a knife," or "I was trying to leave and he wouldn't let me go." Don't assume anything without an investigation. Look and listen for any corroborating evidence of statements regarding self-defense. Document the demeanor of the individual without personal judgment. Anxieties, nervousness, anger, or depression are all emotions that should be noted in a subsequent report in addition to any statements made. The individual who claims, "I had no choice, she was going to leave" or "He was fooling around on me" is trying to justify the abuse. This is not suggestive of self-defense! When severe injury has been inflicted, even the self-defending individual will most likely warrant an arrest. If both parties seemed involved in the violence, consider arresting the primary aggressor and summonsing the other person.

Physical injuries that are readily apparent and indicative of self-defending include scratch and bite marks. The locations of these injuries are important indicators. Scratch marks on a face or chest may suggest that the one who inflicted them was trying to force the person away from them. Bite marks to the arms or neck may have been inflicted in an attempt to free oneself or to get someone off of them. On the other hand, injury to the inside of the forearms often signals an attempt to protect oneself.

Always ask, "Are you okay?" Allow the person to answer in his or her own words. Prompt for clarification without interrupting whenever possible. Any indication that harm was committed should be followed up with an examination by the police officer, regardless of the officer's own feelings about the extent of harm that was committed. Don't assume anything until both parties and all witnesses have been questioned.

False Assumption: The Police Will Find the Victim More Likable Than the Perpetrator.   Often the victims of abuse will be interviewed when they are in crisis. Frustrations, anger, fear, and shame are just some of the emotions that they might be experiencing. The mix of emotions may cause the victim to yell and scream at the individual attempting to intervene. It is conceivable that the victim will be the least "likable" of the individuals in a domestic violence situation. An objective interview will not be based on the sympathies of the interviewer toward any party.

False Assumption: The Woman Is Always the Victim in a Male/Female Dispute. Gender is the most obvious bias in cases of domestic abuse. Reliance on this criterion has caused police officers to be unwilling or unable to assess each individual dispute. The police officer who expects to find female victims and male perpetrators comes to rely on this easy resolution to a difficult situation. He or she may not be aware of this bias, one that will certainly influence investigative actions. Studies indicate approximately 5% to 15% of domestic violence cases involve the male as the victim (Cook, 1997). In jurisdictions where department policy mandates arrest regardless of gender, the statistics are higher.

It should be noted that both biased and nonbiased investigations commonly reach the same conclusion—that women are the most frequent victims of domestic abuse. Of more than 960,000 incidents of violence against a current or former spouse, boyfriend, or girlfriend, approximately 85% of the victims are women (Greenfeld et al., 1998). However, preconceptions disallow alternative possibilities. Purposeful nonbiased identification takes time and effort but is procedurally more fair.

False Assumption: Same-Sex Violence Is Mutual Fighting.   When the perpetrator and victim are the same gender and roughly the same physical size and strength the violence is often misperceived as being mutual (Letellier, 1996). This misconception results in frequent dismissals of victims' complaints in same-gender relationships. Gender, size, and apparent ability to cause harm are all factors that need to be considered during the investigation of domestic violence. External characteristics should not be relied upon for victim identification.

Critics charge that police officers dismiss claims of same-sex battering rather than taking the time to adequately assess the situation. Experts are beginning to acknowledge the dilemma and offer direction on what to do. Attempts should be made to determine the *primary physical aggressor.* Here are some things to consider (Healey, Smith, & O'Sullivan, 1998):

- **Do not assume that the physically larger partner is always the primary aggressor.**
  Care must be taken to question the couple and any witnesses closely before making an arrest.

- **Be aware that bruises may take hours to appear, whereas signs of defensive violence, such as scratching or biting, are immediate.**

  Question the partners separately and determine *how* the visible marks were made and why. Did the victim bite an arm that was holding him or her down, for example?

- **Determine if there has been a history of abuse.**

  Batterers tend to recommit. It is more likely that the perpetrator of violence in the past has perpetrated again. Ask if there has been a recent escalation in the violence and why.

- **Victims may feel free to express their anger about the violence to police.**

  Anger or the expression of anger should not be mistaken for primary aggressive behavior.

- **Do not allow yourself to be provoked into arresting both partners.**

  An angry victim's conduct is not a justification for arresting. Failure to "shut up" is not criminal behavior or indicative of who the batterer is.

- **Determine who the *initial* aggressor was—who started it?**

  When both parties exhibit signs of injury, consider the possibility of self-defense and examine the relative level of injury or force involved. Purposeful interviewing seeks to identify the person who is responsible for the altercation. The source of the argument may not necessarily be evident from that one incident. Question both parties about the history of abuse that describes a battering cycle. Find out which person is the one in *fear,* and why.

Risk Assessment

In order to make a decision on how best to protect and advise the victim, a risk assessment is necessary. This evaluation is the second objective for interviewing in the domestic violence case. High-risk offenders assault their partners an average of 60 times per year, whereas the partners of battered women in the general population engaged in an average of five assaults a year (Straus, 1993). These violent offenders are those known to have initiated three or more instances of violence in the previous year, threatened their partner with a weapon in hand, verbally threatened to kill the partner, or caused the need for medical treatment by the victim (regardless of whether it was obtained). Other high-risk factors include excessive drinking or drug use and threats or actual killing or injuring of a pet. There is a realistic concern for the life and safety of domestic violence victims; their home is at least four times more likely to be the scene of a homicide (MacNeil et al., 1998). Here are some questions to ask that help to determine if the victim is in danger:

- **Does the suspect believe that the victim is attempting to end the relationship?**

  This is the time when the majority of killings take place. The most dangerous period for the victim is when the batterer realizes that this time the victim is really leaving.

- **Does the suspect possess weapons?**

  This question must be asked in all situations. Guns, knives, nunchaks, and any other weapons possessed by the perpetrator or within the home should be confiscated, even if they have not been used against the victim. If they have been used before, the danger for the victim is much greater. Threats to use weapons, even if they are not used, are just as serious as when a weapon has been displayed. A search for weapons can be made without a search warrant on the consent of one person, over the objection of the other, if the residence is shared.

- **Does the suspect abuse alcohol or drugs?**

  The batterer who abuses substances is an increased risk to commit a dangerous or lethal assault. There is some evidence that the perpetrator drinks excessively in order to commit the domestic abuse with justification (Gelles, 1974).

- **Were threats made?**

  Threats to kill the victim or the children or to commit suicide must be taken seriously. Batterers do not commonly kill themselves without first attempting to kill at least one family member. The threat or actual killing of an animal is meant to convince victims that they are likely to be the next targets.

- **Has the suspect committed any previous sexual assaults against the victim?**

  Previous sexual assaults are indicators that the batterer is almost twice as likely to commit a dangerous act of violence against the victim. As of July 1993, marital rape became a crime in all 50 states, the District of Columbia, and on federal lands (National Clearinghouse on Marital and Date Rape, 1996). There is no sexual privilege attached to a wedding vow. If an allegation of rape is made, it should be treated the same as any report of stranger rape. Evidence of the crime must be documented and collected. If possible a specialist in rape investigation should be contacted.

- **Has the suspect been following the victim?**

  Stalking, reading of mail, listening in on phone calls, or other acts of surveillance should be considered an implied threat. Approximately 8% of women and 2% of men have been stalked at some point in their lives (National Institute of Justice, 1998). A former or current intimate partner is the most frequent stalker of women, often while the relationship is still intact, according to the National Institute of Justice report.

- **What is the frequency and severity of the violence?**

  The best predictor of future violence is past violence. Frequent violence (two to three times per month or more) or severe violence (requiring hospitalization) places the victim in a high-risk category. Has the frequency or severity changed? What has happened that contributes to this difference? Has the abuse moved to a public forum?

- **Is the batterer depressed or suffering from mental health problems?**

  Some mental health problems are linked to increased propensity to commit a lethal assault. Look for delusional fears and for severe depression. Research cautions against using mental health issues as excuses for battering, however. An extremely small number of batterers are believed to suffer from mental health problems that cause the abuse.

Evidence Gathering

A third reason to interview in cases of domestic violence is for the identification of evidence so that it might be collected and preserved. This includes both physical and verbal evidence. Frequently the victim will recant or refuse to testify in cases of domestic violence. Presumptive and no-drop prosecution policies that have recently become the trend across the nation require that the case proceed without victim cooperation. Without evidence collection, the prosecution is difficult to nonexistent.

Allegations that a victim had been beaten should be followed up with a visual examination (if appropriate) by the police officer. Bruises, redness, and any indication that violence was perpetrated against a victim must be documented first in the officers' report and followed up with photographs. If victim cooperation is achieved, photographs should be retaken after 24 hours. The level of cooperation may drop after the crisis is over, depending on the relationship of the victim to the offender, the severity of the offense, and the length of the relationship. Do not rely on returning to the scene to gather evidence without a warrant for reentry.

Photographs of the suspect should be taken as well. When allegations of rape have been made the evidence on the perpetrator may be as important as on the victim. Scratch and bite marks that indicate that the partner was struggling rather than consensual give credibility to the victim. While it may be obvious for cases involving children or elders, even forceful sex with an adult partner can leave marks of evidentiary value. Photographing the perpetrator documents the presence or lack of injury in addition to demeanor and appearance at the time of arrest. The suspect may look very different in a suit and tie at the court appearance!

If a person says that he or she was pushed, ask if the person hit anything or fell as a result. Is there soreness or tenderness associated with the fall? Did furniture or household articles get knocked over? Any damage to the individual or to the household caused by a "push" is indicative of the use of force and should be documented.

Do not allow the individual to minimize injuries, a common problem in abuse reporting. If the person states that she or he was choked, ask for a description of that—remember that people choke on bones when swallowed; hands placed around the neck with pressure applied is a *strangulation* attempt. Examine the complainant for any redness or marks that may be identified as finger or cord marks on the throat. Remember that bruises may not be visible for days, so a follow-up should take place with any complaints of injury if possible.

Common household items are frequently used as weapons in domestic violence assaults. If an allegation involves assault with any weapon, it should be confiscated. Items to consider include a hairbrush, cord, telephone, telephone wire, kitchen appliance, or baseball bat. Listen to the victim during the recollection of events for clues on what to look for. Hair clumps, torn clothing, and other items indicative of a struggle should be seized. Physical evidence can bolster any case and is paramount in cases of domestic violence. If sheets, blankets, pillows, or sections of furniture are blood stained, photograph and remove them for evidence.

The allegation of domestic violence assault carries with it responsibilities and duties for the police officer. Although the reports must be taken seriously, they can never be used as an excuse to conduct an illegal search of a person's home. If both people involved in a domestic dispute request that the police leave the premises, they must do so, unless probable cause exists that a crime has been committed. Searches to uncover evidence will violate the expectation of privacy guaranteed under the Fourth Amendment of the Constitution of the United States.

The *excited utterance* takes on immense importance in allegations of domestic violence. Due to its being an exception to the hearsay rule, police officers and others may testify to spontaneous statements from victims, witnesses, and perpetrators under certain circumstances. In some cases the statements may be admitted even when the person chooses not to testify against the perpetrator and regardless of whether the declarant is available to give courtroom testimony.[5] Typically this includes statements about physical condition, pain, intent, plan, motive, design, mental feeling, or emotion. The spontaneity of the statement should be documented in the officers' report in addition to any apparent nervousness, pain, or anxiety that is evidenced. The excited utterance is defined as:

> A statement relating to a startling event or condition that is made while the declarant was under the stress of excitement caused by the event or condition (Black, 1990).

When responding to the scene of a domestic violence assault, it is the duty of the officers present to determine if *any* crime has been committed. This means that evidence of child or elder abuse should not be ignored simply because it was not the reason for the initial call. When interviewing children as witnesses of abuse, questions should also be asked to determine if they have been victimized. Are there signs of neglect? Police officers are required by law in every state to file with a designated agency whenever they have reason to believe that a child, disabled person, or elder is being abused or neglected. Don't forget that responsibility exists.

Outcome Determination

The fourth reason for purposeful interviewing in the domestic violence case is to determine the appropriate police action. In the midst of chaos it is tempting to arrest everyone involved and to let the courts sort it out. Even easier, the responding officers may leave the scene on the "promise" that no one will cause any further problems. Neither of these outcomes is desirable for the majority of family abuse cases.

Police officers, like most people, seek out the most comfortable solution to the situation in which they are involved. Disputes are notoriously uncomfortable, and few welcome the opportunity to respond. Armed with this understanding, police officers can prepare themselves for the chaos that may exist (or develop) and logically seek the facts that will assist them in making the outcome determination. The final decision must be based on the facts at hand rather than the emotions of the police officers involved.

1. Know what is expected. Statutes governing police action in domestic assault cases vary from state to state. Police officers must learn what is expected in the particular jurisdiction that they patrol. Knowing the *required* response according to the particular jurisdiction is a priority. Some states require that an individual be arrested for any

---

[5]See *Commonwealth v. Whelton,* 1998; Federal Rule of Evidence 803 (3).

domestic assault. Others express a preference for an arrest. Police department policy may particularly encourage that the responding officers arrest the perpetrator. Violation of a civil restraining or no-contact order often requires that an officer make an arrest.

Police officers have the power to arrest for all felony crimes for which they have established probable cause. Some misdemeanor crimes carry with them the statutory power of arrest. This is true for many crimes that are committed in a domestic relationship. As stated earlier in this chapter, there is little consistency from state to state on how police officers must respond to cases of domestic violence. State and federal policies strongly discourage mutual arrest. It is imperative that officers know the legal requirements and standards of the particular jurisdiction.

2. Establish probable cause.   A person cannot be arrested for failure to obey an officer's request, acting angry, or being uncooperative. No persons can be legally deprived of their liberty without probable cause that a crime has been committed and that they are the ones most likely to have committed the crime. Regardless of any legal mandate or preference for officers to make an arrest in cases of domestic violence, none can occur unless probable cause exists. Can an arrest be made simply because a person states that he or she was assaulted by an intimate? In some jurisdictions the answer is yes—a mere allegation may rise to the level of probable cause in the absence of conflicting information or evidence if the party is trustworthy and reliable. This means that for officers to rely on mere allegation they must attempt to determine the trustworthiness and reliability of the allegation. Probable cause for an arrest must be articulable. It must be present at the time that the arrest is being made or based on events that have already occurred.

Probable cause to arrest exists where facts and circumstances within officers' knowledge and of which they had reasonably trustworthy information are sufficient in themselves to warrant a person of reasonable cause in the belief that an offense has been or is being committed. It is not necessary that the officer possess knowledge of facts sufficient to establish guilt, but more than mere suspicion is required.

How is probable cause determined? First, determine what crime has been committed. In the domestic violence situation the most typical violation is an assault. Statutes can be broken into elements; have they all been satisfied? Behavior can be reprehensible, but unless it constitutes a violation of all of the elements of the crime it is not criminal behavior. Absent criminal behavior, police officers have no power to arrest. An example would be someone who is yelling and screaming. These actions could be annoying or even taunting to the officer, but this is not criminal and therefore cannot be justification for an arrest.

INTERVIEW METHODS

In order to accomplish the four goals of interviewing for the domestic dispute, separate the parties immediately and question them individually after a brief cooling-down period. The interviews should be outside the hearing range of the other with equal levels of respect afforded to each. For crimes involving intimates the process may be a long one. Sufficient time to assess the situation involves skillful interviewing and the desire to be thorough. There is no quick resolution to a domestic dispute. Everyone present should be questioned, including children and elders, who may have been direct victims or have witnessed the abuse.

Eyewitness reports of crimes are notoriously incomplete and sometimes unreliable. Witnesses see part of what actually happens and recognize only part of what they see. Frightened, angry, or emotionally upset witnesses perceive only a fraction of what they might otherwise observe under normal circumstances. Witnesses can be affected by physical or emotional factors that may influence the validity of their information. A knowledgeable investigator will realize that emotional factors can cause a witness to lie, to become uncooperative or forgetful, or to give prejudicial information. Skilled investigators know that individuals asked to give information about a situation in which they are involved may not give a fully accurate statement. This is particularly true if the victim has a personal stake in the outcome of the event.

The victim of domestic violence may be feeling a sense of betrayal and a loss of trust. This is not unusual since the perpetrator is a person that the victim may have trusted and even been intimate with in the past. Don't expect that she or he will immediately offer information to the investigator. Guilt, shame, disbelief, and anger are all emotional repercussions to an assault from a family member and loved one. What can be done to improve the interview process given these complex emotions? Most important, the interviewing officer should establish a cooperative relationship with the person being interviewed; this is referred to as "rapport." All persons must be allowed to give their side of the story, even if it is not what the officer wants to hear. This means that the witness should not be interrupted unnecessarily or repeatedly until the story is complete. Distractions should be kept to a minimum. This may be difficult during the dispute call, but it is worth the effort to find a quiet corner in order to talk. Specific techniques have been developed to aid the police officer in obtaining accurate and complete information about the event that is being investigated.

Look for the physical signs that someone has been victimized. External signs include bruising, impression marks, swelling of the neck, raspy voice, sore throat, bleeding from the mouth, scratches or cuts, and bite marks. Follow up on the visual cues by asking the victim specifically about what has been noticed. Ask the person if she or he has difficulty breathing, feels light-headed, or has any difficulty in swallowing. Inquire about what was being said at the time of the assault.

Traditional Approach to Interviewing

Most police officers are taught to rely on the "Five W's and How." These are—who, what, where, when, why, and how. Reminiscent of the *just the facts* approach, this traditional method of interviewing should describe the information that must be documented rather than how the interview should be conducted. While talking to the witnesses it is important to keep these traditional questions in mind; they can serve as reminders of what has not been asked. Eliciting information through a precrime and postcrime phase of questioning may assist in establishing a motive. What happened before the violence erupted? Is this indicative of a pattern of conduct? Document those statements that are suggestive of a power differential within the relationship.

Common precipitating events include actions of the victim that seem contrary to "directions" or "instructions" from the perpetrator or signify overrigid expectations. Statements from the victim often minimize the event and project self-blame. Examples:

"I should have known better . . . . "

"The meal was cold, it is all my fault."

"He told me to be ready at 6:00 P.M. but I was late."

"I just didn't want to go shopping tonight, but we go every Thursday."

"He didn't mean to . . . ."

Even when you readily identify the perpetrator, talk to him or her! Acknowledge the suspect's frustrations and anger. Allow the suspect to tell his or her side of the story. Be sure to document any spontaneous admission. Miranda warnings are required when a person has been arrested or otherwise deprived of his or her liberty. It is not necessary in the majority of domestic disputes to give these warnings at the onset of the response. Take the time to develop sufficient probable cause before making that final determination and placing someone in custody if the situation warrants it.

The traditional approach to eliciting information may be a good place to start an interview and it does provide a reminder prior to closing of the facts that must be established. An officer who relies on this approach must take care that she or he does not slide into the automated and dispassionate interview style that we tend to think is associated with the traditional interview. Providing empathy and active listening skills will supplement this basic traditional method and greatly improve the likelihood of good interview results. However, the traditional method is only one of several ways to approach the domestic violence interview. The remainder of this chapter offers other methods of obtaining the requisite information.

Behavioral Approach

The behavioral approach is considered a more comprehensive plan of interviewing. Three basic phases to the interview process are preparation, establishment of the psychological content, and the actual questioning. Obtaining information for the dispatcher en route to the scene or from officers already on site satisfies the preparation stage. If the interview is staged at the convenience of the officer rather than from crisis intervention, then additional preparation would involve contacting agencies that have previously been involved. Records checks and witness interviews would also be considered part of the first phase.

The second phase contains the most important element called *rapport,* which describes the relationship that is established between the interviewer and interviewee. This relationship can be constructed or destroyed within seconds when the responder shows distaste, distrust, or condemnation of the person who is confronted. Evidenced bias such as asking the woman what she did to provoke an attack, or assuming that a blood-soaked man was the abuser are examples of what breaks down the rapport and inhibits information that would benefit the officer and facilitate a resolution. Victim cooperation is consistently tied to officers' attitudes. Look at the evidence and question both parties without prejudice. If victims do not believe that they will get fair treatment from the officer, then they don't bother trying. If they won't give information, then you can't proceed.

Since the establishment of rapport has been identified as an important component for successful interviewing, it bears some discussion here.

> Rapport is a state of mind, and although most people recognize it, few can define it. Words such as empathy, liking, and comfort come close . . . (its) elusiveness stems from the requirement that two individuals achieve a state of harmony through an informal process that has no rules . . . the responsibility for achieving this harmonious state lies exclusively with the interviewer . . . (Ramsey-Klawsnik, 1993, p. 11).

Experts suggest that techniques to put the interviewee at ease or using flattery can facilitate rapport building. It is not difficult to put someone at ease when they are nervous at being interviewed. Offer a glass of water or a cup of coffee; the interviewer may allow the interviewee to have a cigarette. It is up to the person doing the interview to determine the extent to which he or she wants to put the individual at ease. However, if you start an interview hard-handed it is almost impossible to deescalate the situation. Even if an interrogation is likely to follow the interview, start slowly and without intimidating the individual. Although flattery is suggested as a rapport-building option, it should be exercised with caution. A member of the opposite gender should never comment on the appearance of the person being interviewed. Acceptable flattery would more likely be about the "nice home," or "good cup of coffee" that is being served.

The third phase is the actual questioning. Questions should be simple and clear. This does not suggest that the interviewer would "talk down" to the victim, but rather that he or she does not assume knowledge. The behavioral approach suggests that the victim or witness be allowed to tell the entire story and that the police officer will clarify what has been said. Any gaps in the information or statements that seem contradictory would be questioned. It is important to listen to what is said, and to document how it is said. Emotional outbursts and voice inflections may offer clues as to what is really happening. The interviewer should follow up by asking for more detail or clarification.

In this behavioral approach the investigator should also give clues that indicate he or she is actively listening. A nod of the head, an appropriate smile, a curious expression, or leaning slightly forward indicates some response to the interviewee. Purposeful silence or change in the tone of voice signals to the interviewee that you have expectations of him or her. Avoid rolling of the eyes, but do not hesitate to have eye contact with those you believe are guilty of wrongdoing.

The Kinesic Interview Approach

The study of nonverbal communication through body movements is referred to as kinesics. The *kinesic interviewing technique* and *neurolinguistic programming (NLP)* are two examples that are based on the principle that body language is unconscious and difficult to control. Central to the approach is reliance on the composition of communication: more than half of all personal exchange of information (55%) is achieved through nonverbal body language; 7% of communication is through verbal content; and approximately 38% is through voice tone or inflection (Bureau of Alcohol, Tobacco, and Firearms, 1997). A person who has committed an illegal act, domestic violence or otherwise, knows that it is not in his or her best interest to give information to the police. It is common for a perpetrator to offer deceptive information, rather than none at all. The guilty person and the innocent person share a common trait: they will deny having committed the crime! Kinesic techniques seek to distinguish between the liar and the truth teller by observing body motion.

The two-phase kinesic interviewing technique was introduced during the 1970s as a practical approach to interviewing (Link & Foster, 1985). It is most frequently associated with interrogation. Through the *detection* and *interrogation* phases, investigators are taught to discover self-initiated verbal and nonverbal behaviors that may indicate untruthfulness. Posture that is inconsistent with the verbal account being offered by the suspect acts as a red flag that deception may be occurring. A guilty person may shake his or her head yes and state that he or she was not at the scene, for example. When a guilty person is asked, "Did you hit your wife?" the involuntary response may be a look of contempt, such as a smirk. Dryness of mouth and restlessness exhibited from the person being interviewed may indicate that he or she is being untruthful. Liars often ask to have the pivotal question repeated while fidgeting or changing body position. The crossing of arms over the chest is considered protective posturing, which may indicate that the person is attempting to distance himself or herself from the interviewer. When deception is suspected it allows the investigator to redirect questions and to confront the witness.

Visual sensory patterns or eye movements provide one window to the person being interviewed, according to the neurolinguistic programming approach. Advocates suggest that 60% of the population is visually dominant; 20% is auditory; and the remaining 15% kinesthetic (Bureau of Alcohol, Tobacco, and Firearms, 1997). When an interviewer determines the sensory system of the person who is being interviewed, it is possible to establish rapport more easily and therefore facilitate communication in general. Eye patterns associated with emotion or sensation offer clues to how the person is feeling during the interview. Some common clues include looking downward; looking downward to the right; eyes closing; and eyes fluttering or blinking (Bard, 1970).

Reid Technique of Interview and Interrogation

The Reid behavior analysis interview has been used since 1948 (Rabon, 1992). During an interview designed to be neutral and without blame, it prepares the investigator to determine if the individual is telling the truth or withholding relevant information. The questions are developed to elicit verbal and nonverbal responses to show inconsistencies, and profile the subject against established models of behavior.

Referred to as affective or humanistic interviewing, today's version has been updated to purposely promote ethical interviewing practices in policing and law enforcement (Reid & Associates, 1999). Central to the method of interviewing is the recognition of the psychological needs of the persons being interviewed and their being satisfied. Successful interviewing occurs when the individual feels physically safe, is allowed to save face, and is able to maintain control by answering questions rather than submitting to the interviewer. Recognizing that people lie and rationalize in order to preserve their self-image the investigator looks for signs of deception and makes a conscious effort to redirect the statement based on those observations.

Personal spacing is a consideration in the affective approach. Care is taken to place the interviewee in relation to the desired effect. Keeping a distance of six feet from the person being interviewed maintains personal space and increases personal comfort. Bringing the interview closer, from two to four feet apart, allows some sense of intimacy when appropriate. The touch of an arm, for example, may take place at the closer distance. The intimate location is at two feet. It can be the most reassuring or the most stressful depending on the intent of the interviewer.

Cognitive Interview

In 1985 the National Institute of Justice announced a new tool for police officers designed to improve the quality of information provided by eyewitnesses to crime (Hess, 1997). Since that time the cognitive interview has gained credibility in the field. The original method consisted of four memory-jogging techniques that are grouped to improve retrieval from the stored memory through the identification of cues and to encourage additional retrieval using unknown memory paths. The cue retrieval portion of the interview involves incident reconstruction; associations with feelings and spontaneous descriptions are encouraged to complete the memory of the incident. It has long been recognized that we associate with smells, tastes, sounds, and feelings the things that occur around us. These first two stages utilize the senses to enhance memory. The second technique grouping involves recalling events through different references such as reverse order and by imagining that the witness was someone else watching the event. It is believed that events recalled from different perspectives might actually be more truthful as witnesses report what happened rather than what the witness may have expected to happen due to life experiences. In the first major study of the cognitive interview method, it was shown to have elicited significantly more correct information than both hypnosis and standard interview methods (Hess, 1997).

INTERVIEWING CHILDREN

Children will be present in homes where police officers intervene in domestic violence calls approximately half of the time. They may be experiencing guilt for failure to intervene in the dispute, anxiety because of the battering, or fear of being abandoned by one or both of the parents. Some of these children may experience stress-related disorders, cognitive problems, or language problems. Care must be taken not to intimidate or frighten the children further. They may have been injured when household items were thrown or weapons used.

When children are interviewed as witnesses in a domestic dispute, a simple traditional interview method is typically adequate to meet the needs of both the children and the police officer. The purposes of interviewing might include the need to assess the level of danger to the children and the need for protective custody; to determine whether abuse occurred; to identify the abuser; and to gather evidence of an assault that they may have witnessed in their home.

Children should always be questioned apart from the parents. The parent should not be within the child's line of vision or earshot. A brief introduction to the child should be made in addition to the reasons why he or she is being interviewed. The language should be kept age appropriate. A few quick questions to determine the level of understanding are helpful. If the child is young, ask if he or she can read or write. Does the child know the alphabet? While the interview is being conducted, watch the child for any signs that he or she is injured. Does the child wince or favor an arm? Look for obvious bruising and signs of abuse. Adolescents sometimes adopt a "tough" façade and may not volunteer information on injuries.

Children need direct questioning. Don't beat around the bush or put words into the child's mouth. Be honest with children—they know when someone is lying to them. Don't make any promises about the parents or what will happen to them that is not accurate and honest. If a youngster wants information that you do not think is in his or her best interest, tell the child that you are unable to answer the question at this time.

Cognitive Method

During the early 1990s the cognitive method was evaluated as a technique that might prove effective for use with child victims (Yeschke, 1997). Child witnesses suffered from a lack of credibility and their statements were frequently viewed as unreliable. In an effort to bolster the effectiveness of those questioning children, the memory enhancement cognitive method was specifically redesigned. It consists of three phases; the first, and most important, phase is rapport development. Interviewers are taught how to establish a relationship with the child being interviewed without being patronizing or coercive. Phase two involves techniques to elicit as complete a narrative account of the event as possible from the child witness. The third phase is used to clarify events from the narrative and to build upon the statement using backward-order recall, alphabet search, speech characteristics, and recall from a new perspective.

Forensic Interviewing

When primary victimization is suspected, a child specialist should conduct an in-depth interview. The interviewer may be a police officer, a child psychologist, or a certified forensic interviewer. The most recent trend is toward the use of a forensic interviewer who can establish a nonbiased position as neither an advocate for the child nor an employee of the court. Since 1978, the Center for Child Protection has used the forensic model for interviews with child victims of abuse (Davies et al., 1996). This model involves a multidiscipline approach that is meant to reduce the number of interviews that may ultimately contaminate the child's testimony through repetitive questioning. It involves a standardized procedure that documents the affect or nonverbal report of the child as well as the verbal account. Often this is accomplished through videotaping of the victim during the interview.

The interviewers are familiar with the developmental stages of children and the frustrations of trustworthiness that accompany the child's statement due to age. Using a team approach, the protocol heightens the credibility of the child's statement through careful phrasing of questions so as not to be leading or judgmental. The *forensic interview* is a useful approach to gather information about suspected victimization to young children. Typically employed when sexual abuse is suspected, it is useful to obtain the details on physical abuse also.

Forensic interviewing is a structured format that requires planning and precision. A detailed history of abuse or suspected abuse is obtained prior to the interview. An interview room is specifically designed for the comfort of the designated age group of the interviewee. Direct questioning is necessary to elicit details of an assault, but care is taken not to be suggestive in the form of questioning.

The vast majority of children do tell the truth when questioned; however, their recollection may be inaccurate or incomplete depending on their age, both developmental and cognitive. A child under the age of 10 may not have a sense of time sequence, for example. It would be perfectly normal for a young child to have difficulty in ordering offenses accurately. A child over the age of two may not be able to verbalize single offenses committed against him or her, but may be able to act it out. Some obvious fantasy may be mixed into an account—this does not mean that the entire story is fantasy. An interviewer of children must be trained to recognize inconsistencies and understand what the child is saying.

INTERVIEWING THE ELDER VICTIM

Interviewing the elder victim of domestic violence can present unique challenges to policing. Police officers may sense aggression, agitation, and/or depression; these could represent signs of abuse or be symptoms of a form of dementia such as Alzheimer's disease. The officer may need to rule out organic causes for these symptoms when investigating an abuse situation. At the same time, dementia is a factor that also puts the elder at higher risk for abuse. Approximately 50% to 75% of patients who reside in nursing and foster homes suffer from dementia. Experts believe it is the leading risk factor for elder maltreatment (Weinberg & Wei, 1995). Diminished capacity does not mean that abuse cannot occur; it can and does. Overmedication, misuse of restraints, and denial of adequate food are all examples of abuse to the elderly.

Empowering the elder is helpful to the victim and to the officer as well. The interviewer may accomplish this by giving victims as much control as possible, such as refraining from taking notes before asking permission to do so in the interview and asking where they would want the interviewer to sit (Ramsey-Klawsnik, 1993). Communicating respect by referring to the victim by his or her title—Miss, Mrs., Mr., or Dr., for example—can be helpful during the rapport-building portion of the interview. At all times, dignity of the elder should be maintained; the interviewer should go slowly and use language that is easily understood, but not patronizing.

According to the Police Executive Research Forum Module III-10, the following guidelines are suggested for interviews with elders (Nerenberg, 1993).

1. Investigations should be coordinated with adult protective services or the ombudsman when possible, to avoid multiple interviews.
2. Joint interviews with public health nurses or others treating medical conditions may assist agencies when appropriate.
3. Interviewers should attempt to establish rapport with the person being interviewed.
4. Whenever possible, interviewers should use audio-video technology.
5. Officers should respect confidentiality of all parties whenever possible.
6. Interviewers should avoid disclosure of case information to any parties involved in the alleged offense to prevent contamination or collusion.
7. Interviewers should ask nonleading, general questions.
8. Officers should ask all witnesses to identify others who have relevant information and tell how those others may be contacted. Interviewers will need to identify the victim's doctor, conservator, attorney, social worker, and any agencies that provide services to the victim.

Elders have a number of issues that differ from other populations who are victimized by domestic violence. They are often dependent on abusers for care or companionship. The possibility of the elder being denied access to the perpetrator may be worse to them than the abuse they have suffered. Sensitivity to these issues is paramount for any officer attempting to interview an elder citizen suspected of being victimized through domestic violence.

CONCLUSION

Violence against intimates and family members is a pervasive problem in our society. Police officers are the first line of defense for the victims of abuse within their own households. Following legislative directives and department policies, approximately 40% of the violence that police officers encounter will be domestic. We need to arm these officials with knowledge on the subject of domestic violence in addition to the interviewing skills. Training in purposeful interviewing is goal oriented—the officer knows why he or she is interviewing rather than blindly asking questions. There is a larger picture when domestic violence is alleged, one that is far too complex for simplistic police response.

The skills essential to successful interviewing include patience and tact, respect, control of one's personal feelings, and the ability to encourage uncooperative witnesses by asking appropriate questions. In situations where it is difficult to obtain the information necessary to make a determination, the officers can apply techniques from various approaches to interviewing. At all times the interview should be conducted to corroborate what the officer sees and hears at the scene.

REFERENCES

BARD, M. (1970). *Training police as specialists in family crisis intervention* (NCJ 000050). Washington, DC: U.S. Department of Justice.

BLACK, H.C. (1990). *Black's law dictionary.* St. Paul, MN: West Publishing Co.

BUREAU OF ALCOHOL, TOBACCO, AND FIREARMS. (1997). *Advanced analytical interviewing program.* Washington, DC: U.S. Department of the Treasury, Bureau of Alcohol, Tobacco, and Firearms.

BUZAWA, E., & AUSTIN, T. (1993). Determining police response to domestic violence victims. *American Behavioral Scientist, 36*(5), 610–623.

COLEMAN, V.E. (1996). Lesbian battering: The relationship between personality and the perpetration of violence. In L.K. Hamberger & C. Renzetti (Eds), *Domestic partner abuse,* (pp. 77–101). New York: Springer Publishing Company, Inc.

COOK, P.W. (1997). *Abused men: The hidden side of domestic violence.* Westport, CT: Praeger.

DAVIES, D., COLE, J., ALBERTA, G., MCCULLOCH, U., ALLEN, K., & KEKEVIAN, H. (1996). Model for conducting forensic interviews with child victims of abuse. *Child Maltreatment, 1*(3), 189–197.

GELLES, R. (1974). *The violent home: A study of physical aggression between husbands and wives.* Beverly Hills, CA: Sage.

GREENFELD, L., RAND, M., CRAVEN, D., KLAUS, P., PERKINS, C., RINGEL, C., WARCHOL, G., MASTON, C., & FOX, J. (1998). *Violence by intimates: Analysis of crimes by current or former spouses, boyfriends, and girlfriends.* (NCJ 167237). Washington, DC: U.S. Department of Justice.

HATTENDORF, J., & TOLLERUND, T.R. (1997). Domestic violence: Counseling strategies that minimize the impact of secondary victimization. *Perspectives in Psychiatric Care, 33*(1), 10–14.

HEALEY, K., SMITH, C., & O'SULLIVAN, C. (1998). *Batterer intervention: Program approaches and criminal justice strategies* (NCJ 168638). Washington, DC: U.S. Department of Justice.

HESS, J. (1997). *Interviewing and interrogation for law enforcement.* Cincinnati, OH: Anderson Publishing Co.

ISLAND, D., & LETELLIER, P. (1991). *Men who beat the men who love them: Battered gay men and domestic violence.* Binghamton, NY: Harrington Park Press.

JOHNSON, B., LI, D., & WEBSDALE, N. (1998). Florida mortality review project: Executive summary. *Legal interventions in family violence: Research findings and policy implications.* (NCJ 171666). Washington, DC: U.S. Department of Justice.

KELLEY, B.T., THORNBERRY, T.P., & SMITH, C.A. (1997). Juvenile justice bulletin. *In the wake of childhood maltreatment.* Washington, DC: U.S. Department of Justice.

LETELLIER, P. (1996). Gay and bisexual male domestic violence victimization: Challenges to feminist theory. In L.K. Hamberger & C. Renzetti (Eds.), *Domestic partner abuse.* (pp. 1–22). New York, NY: Springer Publishing Company.

LINK, F.C., & FOSTER, G.D. (1985). *The kinesic interview technique* (2nd ed.). Atlanta, GA: Interrotec Associates.

MACNEIL, R.L., D'ERRICO, J.A., OUYANG, H., BERRY, J., STRAYHORN, C., & SOMERMAN, M.J. (1998). Isolation of murine cementoblasts: Unique cells or uniquely-positioned osteoblasts? *European Journal of Oral Sciences, 106*(Suppl.), 1350–1356.

MOFFITT, T.E., & CASPI, A. (1999). *Findings about partner violence from the Dunedin multidisciplinary health and development study* (NCJ 170018). Washington, DC: U.S. Department of Justice.

NATIONAL CENTER FOR INJURY PREVENTION AND CONTROL. (1998). *The co-occurrence of intimate partner violence against mothers and abuse of children.* Atlanta, GA: Violence Prevention Centers for Disease Control and Prevention.

NATIONAL CLEARINGHOUSE ON MARITAL & DATE RAPE. (1996). *State law chart.* Berkeley, CA: Author.

NATIONAL COALITION OF ANTI-VIOLENCE PROGRAMS. (1998). *Annual report on lesbian, gay, bisexual, transgender domestic violence.* New York, NY: Author.

NATIONAL INSTITUTE OF JUSTICE. (1998). *Stalking and domestic violence: The third annual report to congress under the violence against women act* (NCJ 172204). Washington, DC: U.S. Department of Justice.

NERENBERG, L. (1993). Improving police response to domestic elder abuse. Washington, DC: Police Executive Research Forum.

RABON, D. (1992). *Interviewing and interrogation.* Durham, NC: Carolina Academic Press.

RAMSEY-KLAWSNIK, H. (1993). Interviewing elders for suspected sexual abuse: Guidelines and techniques. *Journal of Elder Abuse & Neglect, 5*(1), 5–17.

REID, J., & ASSOCIATES, INC. (1999). *Reid behavior analysis interview* [On-line]. Available: www.reid.com

SHERMAN, L., & ROGAN, D. (1992). Policing domestic violence: Experiments and dilemmas. New York, NY: Free Press.

STRAUS, M.A. (1993). Identifying offenders in criminal justice research on domestic assault. *American Behavioral Scientist, 36*(5), 587–599.

STRAUS, M.A. (1996). Domestic violence is a problem for men. In B. LEONE (Ed.), *Domestic violence* (pp. 50–64). San Diego, CA: Greenhaven Press, Inc.

SUGARMAN, D.B., & HOTALING, G.T. (1998). Dating violence: A review of contextual and risk factors. In B. Levy (Ed.), *Dating violence: Young women in danger* (2nd ed., pp. 100–118). Seattle, WA: Seal Press.

WEINBERG, A.D., & WEI, J.Y. (1995). *The early recognition of elder abuse.* Bayside, NY: American Medical Publishing Co., Inc.

YESCHKE, C.L. (1997). *The art of investigative interviewing.* Newton, MA: Butterworth-Heinemann.

7

Victims' Rights Legislation

An Overview

Robert A. Jerin

Historically, the crime victim was an integral part of the criminal justice system (Doerner & Lab, 1998; Jerin & Moriarty, 1998; Wallace, 1998). Although during the initial development of the criminal justice system in the United States, public prosecutions were becoming the norm (Matthews, 1998), the victim was a full partner in prosecuting the offender. Slowly, this partnership disintegrated to the point where the victim became just another piece of evidence. Until recently the criminal justice system has operated in a manner where its own convenience was of utmost importance. Whatever was most efficient for the actors in the criminal justice system in determining "justice" was done. This model led to the offender's constitutional rights being ignored and victims' issues never being considered unless they coincided with the needs of the system.

In the 1960s the system started to change and started to operate under an "offender-rights" mode. Key Supreme Court decisions such as *Mapp v. Ohio, Gideon v. Wainwright, Miranda v. Arizonia, Escobedo v. Illinois,* and *In re Gault* all impacted the operation of the criminal justice system (Kalaher, 1997). The system began to take into consideration the rights of offenders, but still ignored the legitimate concerns of crime victims and how they were being treated by the criminal justice system.

Slowly, since the 1960s, crime victims have been able to establish a presence in the criminal justice process and make the system respond to their needs. National movements during the 1960s such as the civil rights movement, the women's rights movement, the law and order movement, and the social welfare movement all contributed to this change in the crime victim's role in the criminal justice process. Grassroots efforts have sought to bring balance into the criminal justice system, to make it more responsive to the needs of crime victims, and to provide assistance to crime victims; citizens have demanded crime victims also be granted rights. Over the past almost 40 years these efforts have resulted in a wide-ranging assortment of legislative and judicial initiatives on both the federal and state level.

To better understand the changes and important events in the crime victims' movement in the United States since the 1960s an examination of the chronology of the movement is necessary. The following is a timeline of crime victims' initiatives and legal decisions that have impacted the crime victims' movement (Doerner & Lab, 1998; Dussich, 1986; Jerin & Moriarty, 1998; Karmen, 1998; Neff, 1997; National Organization for Victim Assistance, 1990, 1993, 1995, 1998; Sabatun & Edwards, 1995; Marion, 1995). The highlights over the last 35 years include:

1964   Kitty Genovese is murdered in New York City.

- The murder, which occurred in plain view of a number of people who ignored her cries for help, led to a national debate and interest in the plight of crime victims.

The first federal victim compensation program is proposed.

- Senator Ralph Yarborough of Texas introduced legislation to create a national crime victim compensation program based upon similar programs developed in New Zealand and Great Britain. It was never passed.

The Civil Rights Act, Title VII, is passed.

- This act forbids discrimination on the basis of sex, race, national origin, and age. This act becomes the foundation for sexual harassment laws.

1965   California passes the first victims' compensation program.

- Patterned after national programs established in Great Britain and New Zealand, compensation was provided from state-administered funds to victims of violent crimes who met certain eligibility criteria including, but not limited to, burial expenses, medical bills, wage losses, and other crime-related expenses not paid by insurance or other sources.

1967   The President's Commission on Law Enforcement and Administration of Justice is established.

- As a result of growing crime problems and antiwar protests of the 1960s, President Johnson appointed a commission to study crime and the criminal justice system. In its report *The Challenge of Crime in a Free Society,* victims' issues were a component.

The first national victims' survey is conducted.

- Sponsored by the National Opinion Research Center, this survey targeted 10,000 households throughout the country. Its initial findings reported that the *Uniform Crime Reports* underreported serious crime in general by roughly 50%, and specifically found rape to occur four times as often and burglaries three times as often.

1968   The Law Enforcement Assistance Administration (LEAA) is established.

- As a direct result of the President's Crime Commission, LEAA was authorized to provide start-up funds for many victim assistance and victim/witness programs.

Steven Schafer's *The Victim and His Criminal* is published.

- Schafer's book was the beginning of an academic examination of crime victims' issues and concerns with the criminal justice system from a victim's perspective.

1969   New York City establishes the Mayor's Task Force on Child Abuse and Neglect.

- This was the first coordinated statewide effort in the nation to examine the problems associated with the definition, detection, and prosecution of child abuse.

1971   The National Crime Prevention Institute (NCPI) was established.

- With funds provided by LEAA, the NCPI, located at the University of Louisville, provides educational and technical resources for the development of crime prevention programs for both the public and private sectors.

1972   The first victims' assistance programs are created:
              Aid for Victims of Crime, St. Louis, Missouri;
              Bay Area Women against Rape, San Francisco, California; and
              D.C. Rape Crisis Center, Washington, D.C.

- Through grassroots efforts, privately run programs staffed by former victims and their supporters are established to help women who are victims of rape and domestic violence.

1973   The first victim impact statement is created in Fresno County, California, by Chief Probation Officer James Rowland.

- A written statement by the crime victim or his/her survivors is used to provide the judiciary with an objective listing of victim injuries and the impact of the crime prior to sentencing.

The first International Symposium on Victimology is held in Jerusalem.

- This was the first international gathering of academics and professionals working with crime victims to study the science of victimology and the impact of victimization.

1974   The first victim/witness programs are created, with funding from the LEAA.

- District attorneys' offices in Brooklyn, New York, and Milwaukee, Wisconsin, along with seven other district attorneys' offices in collaboration with the National District Attorney's Association establish programs with specific personnel to assist crime victims and witnesses with the court process as an effort to increase cooperation by crime victims and improve prosecution of offenders.

The first law enforcement–based, victim assistance programs are established.

- Ft. Lauderdale, Florida, and Indianapolis, Indiana, set up programs to provide crime victims with greater information on the progress of their cases and to provide services for victims.

The Child Abuse Prevention and Treatment Act is passed by Congress.

- This act established a National Center on Child Abuse and Neglect within the federal government and provided the first federal effort to examine the issues of child abuse and abduction.

Families and Friends of Missing Persons and Violent Crime Victims is established.
- This grassroots organization consisting of survivors and family members of victims became one of the first support organizations for crime victims and their family members.

The first rape shield statute was passed in Michigan.
- This law provides the rape victim with protection from having her past sexual history entered as evidence in criminal trials.

1975   The National Organization for Victim Assistance (NOVA) is established.
- This organization became the first of many umbrella and specialized national advocacy organizations and support groups in the United States and other countries established by citizen activists to expand victim services and increase recognition of victim rights.

The first "Victim Rights Week" is organized by Philadelphia's district attorney.
- This was the first public recognition of the impact of crime victimization and the first call to provide greater services for crime victims and their families.

1977   The first legislation mandating arrest in domestic violence cases is enacted in Oregon.
- This legislative effort changed police practices from treating domestic violence as a "private" matter and required police to arrest the offender if the crime occurred in their presence.

1978   The National Coalition Against Sexual Assault (NCASA) is formed.
- Survivors of sexual assault and their supporters organize to combat sexual violence and to promote services for survivors such as rape crisis hotlines.

The National Coalition Against Domestic Violence (NCADV) is established.
- Survivors of domestic violence and their supporters organize to respond to the needs of battered women and their children, and to create recognition that domestic violence is a crime and should be treated accordingly.

Parents of Murdered Children (POMC) is founded in Cincinnati, Ohio.
- This is a self-help support group of survivors whose children were the victims of a criminal homicide.

1979   The first bill of rights for crime victims is enacted in Wisconsin.
- This was the first legislative granting of "rights" that give crime victims a meaningful role in the criminal justice system. These "rights" can include compensation, restitution, protection from intimidation, notification of proceedings and hearings, participation, assistance and privacy, and security of victim information.

The first "National Victim Rights Week" is organized by NOVA.
- The focus of the week is to provide public awareness of the plight of crime victims and to focus public pressure on governmental officials to enact legislation that will benefit crime victims in their recovery and in the criminal justice system.

MADD (Mothers Against Drunk Driving) is founded.

- After her daughter is killed by a repeat drunk driver, Candy Lightner establishes an organization in California to provide support for survivors of drunk driving accidents and to exert pressure on state and federal legislatures to enact and enforce stricter drunk driving statutes and to raise the drinking age nationwide.

1980   The National Crime Prevention Council (NCPC) is established.

- The goal of this organization is to establish a program to provide a comprehensive and coordinated national crime prevention campaign.

1981   The first presidential-commemorated "National Victims Rights Week" is proclaimed by President Ronald Reagan.

- The "Victims Rights Week" is observed in April each year and seeks to focus attention on the plight of crime victims in the criminal justice system, the need for greater services and "rights" for crime victims, and the need for more services for crime victims on both the state and federal level. It also is intended to honor those who serve crime victims.

A special task force to consider victim issues is recommended by the Attorney General's Task Force on Violent Crime.

- This recommendation resulted in the establishment of the President's Task Force.

A national public awareness campaign on child abduction is launched after the disappearance of young Adam Walsh, later found to have been a murder victim.

1982   The Task Force on Victims of Crime is appointed by President Reagan.

- The findings of the task force are presented in the President's Task Force on Victims of Crime *Final Report*. This report becomes the foundation for victims' rights issues for the next 25 years offering numerous recommendations for legislative action on both the federal and state level, recommendations for criminal justice system agencies, and recommendations for other organizations and the private sector.

The Victim and Witness Protection Act is passed.

- The act seeks to bring "fair treatment standards" to victims and witnesses in the federal criminal justice system and to protect federal victims from intimidation and threats.

The first "Victim Impact Panel" is organized by MADD in Rutland, Massachusetts.

- The object of the panel is to get convicted drunk drivers to hear from victims and survivors of other drunk driving crashes in an effort to change the behavior of the offenders.

The Missing Children Act is passed.

- This act ensures that identifying information on a missing child is promptly entered into the FBI National Crime Information Center (NCIC) computer.

The California "Victim Bill of Rights" state constitutional amendment is passed.

- This initiative is overwhelmingly adopted by voters to provide a state constitutional guarantee of restitution and other statutory reforms for crime victims within California's criminal justice system.

1983 The International Association of Chiefs of Police (IACP) adopts a new Crime Victims Bill of Rights.

- This bill of rights addresses the needs of crime victims for law enforcement officers around the country.

The first National Missing Children's Day is proclaimed by President Reagan.

- In response to the disappearance of a child, Etan Patz, a national observation of the problem of child abduction is established.

The Attorney General's Task Force on Family Violence is established.

- As a response to the President's Task Force *Final Report,* a task force is established to examine the particular characteristics of family violence and to articulate solutions to the problem of domestic violence in the nation.

1984 The Victims of Crime Act (VOCA) is enacted.

- Called the "Marshall Plan of the victims' movement," VOCA is enacted to provide for a federal victims' compensation program for victims of federal crimes, federal subsidies for state victim compensation programs that provide a minimal level of coverage for victims of violence, and funds for local and national crime victim service programs.

The Federal Justice Assistance Act is passed.

- This act provides block grants to states for criminal justice system improvements, including the development of victim/witness assistance programs.

The Missing Children Assistance Act is passed.

- The National Center for Missing and Exploited Children is created to expand services and federal assistance to local community agencies working to prevent child abduction.

The Attorney General's Task Force on Family Violence *Final Report* is published.

- The report makes specific recommendations to increase the public's awareness of the nature of domestic violence and ways of combating it. The report focuses upon recommendations for the justice system and for prevention and awareness of domestic violence on all levels.

In the case of *Thurman v. Torrington,* police are held liable for failing to protect a woman from her abusive husband.

- The Federal District Court recognizes the right of a woman to sue local municipalities and police agencies for failing to protect her in a domestic dispute after repeated requests as a violation of the 14th Amendment's Equal Protection Clause.

1985 The International Declaration on the Rights of Victims of Crime and the Abuse of Power is passed by the United Nations General Assembly.

- This declaration by the United Nations is an effort to focus international attention on the plight of victims of crime and abuse of power throughout the world.

The National Victim Center (NVC) is founded.

- In honor of Sunny von Bulow the NVC is established to promote the rights and needs of violent crime victims, and to educate the American public about the devastating effects of crime on our society.

The Office for Victims of Crime (OVC) is established within the Justice Department.

- The responsibility of this office is to assist federal and state victims' service programs, to encourage the development of new victims' programs, and to push for the improved treatment of victims by criminal justice personnel and other professionals.

The President's Child Safety Partnership is formed.

- A group of citizens from the business community, the private nonprofit community, the government sector, and the private sector is appointed to find answers and solutions to the problems of child victimization.

1986   The Child Abuse Victims Rights Act of 1986 is passed by Congress.

- This law created a civil cause of action for children who suffer personal injuries as a result of being victims of federal sexual exploitation law violations.

1987   The Victims' Constitutional Amendment Network (Victims' CAN) is established.

- VCAN becomes the formal outgrowth of an ad hoc committee at an NVC-hosted meeting, seeking to establish a federal constitutional amendment that will guarantee crime victims' rights.

The President's Child Safety Partnership *Final Report* is published.

- The report details numerous recommendations for change in policies and programs for the private sector along with the child-serving community including parents and the various agencies of the federal, state, and local governments.

The Supreme Court rules on *Booth v. Maryland.*

- The Supreme Court rules that victim impact statements are too inflammatory and cannot be used in capital murder cases.

1988   The Federal Drunk Driving Prevention Act is passed.

- The act required all states to raise their drinking age to 21 or risk losing federal highway funds.

The Anti-Drug Abuse Act is passed.

- VOCA is reauthorized for another 6 years and the cap on the funds is raised to over $100 million.
- State compensation programs are extended to cover victims of domestic violence and drunk driving by amendments to VOCA.

1990   The Crime Control Act of 1990 is passed.

- The act includes the Victims' Rights and Restitution Act. This act strengthens victims' rights in the federal system by establishing the federal Crime Victims' Bill of Rights.
- This act also includes the Victims of Child Abuse Act of 1990, which reforms treatment of child victims. Specifically, it improves investigation and prosecution of child abuse cases, and provides court-appointed special advocate programs, child abuse training programs for judicial personnel and practitioners, and other reforms.

1991   The Civil Rights Act is amended.

   • The amendment allows for compensatory and punitive damages in sexual harassment cases.

   The first International Conference on Campus Rape is held.

   • The conference is held in Orlando, Florida, and focuses on the national problem of sexual assaults and crime on college campuses.

   The International Parental Child Kidnapping Act is passed.

   • This act makes the act of unlawfully removing a child outside the United States a federal felony.

   The Supreme Court rules on *Payne v. Tennessee*.

   • Victim impact evidence at a capital sentencing hearing is deemed constitutional over the previous ruling in *Booth v. Maryland*.

   The Supreme Court rules on *Michigan v. Lucas*.

   • Rape shield laws are deemed constitutional.

   A Congressional Joint Resolution to place crime victims' rights in the U.S. Constitution is put forth.

   • The first proposed constitutional amendment following the recommendations of the 1982 President's Task Force on Victims of Crime is introduced by Representative Ileana Ros-Lehtinen (R-FL).

1992   The spending cap and sunset clause are lifted from the Victims of Crime Act.

   • VOCA is given permanent status and the amount of money in the fund is raised from $150 million to an unlimited amount.

   Maine is the last state to pass a crime victim compensation program.

   • The law goes into effect June 1993, providing crime victims in all 50 states and victims of federal crimes with the opportunity to receive compensation for their injuries.

1994   The Violent Crime Control and Law Enforcement Act is enacted.

   • The Violence Against Women Act is included in the act, which provides for federal relief in civil court for women against their offenders.

   Megan's Law is passed.

   • After the death of Megan Kanka at the hands of a just released two-time convicted sex offender, New Jersey passes a law requiring registration and notification when sex offenders and child abuse offenders are released into the community.

1996   In the Senate, a Victims' Bill of Rights to place crime victims' rights in the U.S. Constitution is put forth.

   • The first version of a Victims' Bill of Rights Constitutional Amendment is introduced into Congress by Senators Fienstein and Kyl.

   The Interstate Stalking Punishment Act of 1996 is passed.

   • The act makes it a federal crime to stalk someone across state lines and defines certain acts as stalking.

The Justice for Victims of Terrorism Act is passed.

- As a reaction to the Oklahoma City bombing, victims of terrorist acts are granted rights to sue in federal court.

The Mandatory Victim's Restitution Act of 1996 is passed.

- The act requires that restitution be ordered in federal cases involving violent crime or in limited other offenses.

1997   The Victim Rights Clarification Act of 1997 (VRCA) is passed.

- The VRCA permitted federal victims to observe all portions of the criminal trial and preserved the victim's right to address the court during sentencing proceedings.

1998   The Office of Victims of Crime publishes *New Directions from the Field: Victims' Rights and Services for the 21st Century.*

- Updating and examining how victims' rights and services have been realized since the 1982 President's Task Force on Victims of Crime *Final Report,* this document is a comprehensive report and set of recommendations on victims' rights and services from and concerning virtually every community involved with crime victims across the nation.

1999   Oregon passes a state constitutional amendment guaranteeing victims' rights.

- Oregon becomes the 32nd state to add an amendment to their constitution providing for the rights of crime victims.

2000   The proposed U.S. Constitutional Amendment guaranteeing victims' rights is shelved.

- Senate sponsors of the victims' rights amendment withdrew the amendment from consideration, conceding it lacked the two-thirds vote needed for approval.

The U.S. Supreme Court rules on *United States v. Morrison et al.*

- The Supreme Court rules as unconstitutional a section of the Violence Against Women Act which provided for a civil remedy in federal court for victims of gender-motivated violence.

An examination of this timeline shows that legislation that gave crime victims access to programs to help them recover from their victimization began slowly with grassroots action, then moved to the states, but exploded once the federal government became committed to establishing programs and providing financial support. California, in 1964, became the first state to provide compensation for crime victims (Schlesinger, 1984). Compensation programs are a type of insurance program. They are set up to cover costs due to victimization that cannot be recovered from the offender or other sources. Compensation payments are made from a state-run fund using public monies or fees assessed against offenders to be paid to victims to help them recover from the financial burdens of their victimization.

As a direct result of the grassroots efforts of the feminist movement beginning in the late 1960s, many victims' programs have emerged. The first private programs set up to help crime victims were rape crisis hotlines and battered women's shelters in the early 1970s. These programs are now found in every state and may be run privately, publicly, or a combination of both.

During the mid-1970s the first victim-witness advocates' offices were started in district attorney's offices. These programs are found on the local, state, and federal level. Although there are no standards for the programs, they offer a multitude of services to crime victims both in and out of court (Jerin, Moriarty, & Gibson, 1995). In the mid to late 1970s, nonprofit organizations such as National Organization for Victim Assistance (NOVA), National Coalition Against Sexual Assault (NCASA), Parents of Murdered Children (POMC), National Coalition Against Domestic Violence (NCADV), and Mothers Against Drunk Driving (MADD) also began. In 1984, the federal government started providing funds to federal victims and assisted state programs with the enactment of the Victims of Crime Act (VOCA; 18 U.S.C. 3013). Today, all 50 states and the federal government cover crime victims with various compensation programs and provide additional funds to nonprofit programs through grants.

The provision of restitution by offenders is another type of legislation that seeks to help the crime victim. Restitution, provided for in all 50 states, is the provision of monies or services to the community or crime victim directly by the offender. Restitution had very little acceptance until the passage of the 1982 federal Victim and Witness Protection Act (Pub.L.No.97-291).

Attempts to guarantee victims' rights in the criminal justice process have been enacted on the federal and state level. On the federal level, the President's Task Force on Victims of Crime (1982) attempted to change the U.S. Constitution by adding a sentence to the sixth Amendment, which would have guaranteed victims the right to be heard throughout their judicial proceedings. This attempt failed; however, today most states have enacted their own constitutional amendment or a state statute that provides for victims' rights or guidelines for the treatment of crime victims by the criminal justice system. Further, the federal government passed in 1982 a crime victims' Bill of Rights as part of the Omnibus Victim and Witness Protection Act which was expanded on in the Crime Control Act: the Victims' Rights and Restitution Act of 1990 (42 U.S.C. 10606).

In 1998, a separate amendment, "The Constitutional Rights for Crime Victims" (Senate Joint Resolution 44), was sent to the full Senate for its consideration. This legislation is modeled after the constitutional amendments that have been used in many states. Although the amendment is in limbo, currently 32 states have enacted constitutional amendments. The rights established in various state constitutions, state statutes, or by federal legislation generally include the following:

- The right to be treated with fairness and with respect for the victim's dignity and privacy.
- The right to be reasonably protected from the accused offender.
- The right to be notified of all court proceedings.
- The right to be present at all public court proceedings related to the offense.
- The right to be heard and submit a statement at all public proceedings relating to the crime.
- The right to confer with attorneys for the government in the case.
- The right to restitution.
- The right to information about the conviction, sentencing, imprisonment, and release of the offender.

CRIME VICTIM AND WITNESS RIGHTS

As a victim or witness of crime, you have certain rights under Virginia's Crime Victim and Witness Rights Act. There are specific steps you must take to receive these rights (see the back of this card for a brief summary). For detailed information, obtain a copy of "An Informational Guide to Virginia's Crime Victim and Witness Rights Act" from your local victim/witness program, commonwealth's attorney's office, police department or sheriff's office, or the Virginia Department of Criminal Justice Services.

As the victim of a crime, you may be entitled to:

Information about:

- protection
- financial assistance and social services, including the Criminal Injuries Compensation Fund (crime victims' compensation)
- address and telephone number confidentiality ❶
- closed preliminary hearing or use of closed-circuit television, if you were the victim of a sexual offense
- separate waiting area during court proceedings
- the right to remain in the courtroom during a criminal trial or proceeding

Assistance in:

- obtaining protection
- obtaining property held by law enforcement agencies
- intercession services with your employer
- obtaining advanced notice of court proceedings ❷
- receiving the services of an interpreter
- preparing a Victim Impact Statement
- seeking restitution

Notification of:

- changes in court dates ❷
- changes in the status of the defendant, if he/she is being held in a jail or a correctional facility ❸
- the opportunity to prepare a written Victim Impact Statement prior to sentencing of a defendant

As the witness to a crime, you may be entitled to:

Information about:

- protection
- address and telephone number confidentiality
- a separate waiting area during court proceedings

Assistance with:

- obtaining protection
- receiving intercession services with your employer
- receiving the services of an interpreter

Steps you need to take to receive confidentiality, notification, or release information:

❶ **Confidentiality:** To request confidentiality, you must file a Request for Confidentiality by Crime Victim Form (DC-301) with the magistrate, court, commonwealth's attorney, or law enforcement agency in the locality where the crime occurred.

❷ **Court Dates:** You must give the commonwealth's attorney your current name, address, and telephone number, in writing, if you wish to be notified in advance of the scheduled court dates for preliminary hearings, trials, sentencing hearings and other proceedings in your case.

❸ **Information about release or status of defendant:** You must give the sheriff, jail superintendent, or Department of Corrections your current name, address, telephone number and defendant name, in writing, if you wish to be notified about the changes in the status of the defendant or inmate.

Contact your local victim/witness program, commonwealth's attorney's office, police department or sheriff's office for further information and assistance.

This brochure is supported in part by Grant #98-A9528AD97 awarded to the Virginia Department of Criminal Justice Services by the Bureau of Justice Assistance, U. S. Department of Justice, through the Edward Byrne Memorial Formula Grant Program. Points of view expressed in this document do not necessarily represent the official position or policies of the U.S. Department of Justice.

FIGURE 7-1   An Illustration of Crime Victim and Witness Rights

Figure 7-1 is an illustration of crime victim and witness rights. In Virginia, as in other states, police officers give this brochure summarizing the rights of the victims.

The final area of victims' rights legislation has been to change or enact new criminal statutes that would treat all crime victims fairly and protect them. The women's movement of the 1960s and 1970s was instrumental in putting this agenda forward. Speaking out against the treatment of women in cases such as rape and domestic violence focused the attention of legislators to mandate greater protection to women and all victims of crime. Victims' supporters in state and federal legislatures have enacted laws embracing crime victims' issues. Crime victim legislation has changed the definition of what constitutes a crime and the sanctions applied to criminals. The crime victims' movement and the women's rights movement in particular have helped enact legislation covering drunk driving, domestic violence, child abuse, sexual assault and rape, elderly victimization, and stalking. The criminalization of domestic violence, rape in marriage, date rape, drunk driving, and stalking is a result of the grassroots movement of women's organizations seeking equal protection from the criminal justice system.

CONCLUSION

Victims' rights legislation recognizes the "right" of all crime victims to be treated equally, to feel safe in their homes, to be protected and respected by the criminal justice system, and to be made whole again if the government fails to protect them from harm. As noted in *New Directions,* "As a society we have made great progress in meeting victims' needs since the 1982 *Final Report* of the President's Task Force on Victims of Crime" (U.S. Department of Justice, 1998, p.429). This progress continues today and into the foreseeable future. New programs for crime victims are being developed and new laws supporting crime victims are being passed. On the federal level, bills currently awaiting legislative action focus on children's rights and rights for victims of hate crimes, stalking, and gender violence, just to name a few. The expansion of "rights" for crime victims will continue until such time as the public believes that they are getting a fair shake from the criminal justice system.

REFERENCES

DOERNER, W.G., & LAB, S.P. (1998). *Victimology,* (2nd ed.). Cincinnati, OH: Anderson Publishing Co.

DUSSICH, J. (1986). The victim assistance center: Its history and typology. In K. Miyazawa & M. Ohya (Eds.), *Victimology in comparative perspective.* Tokyo: Seibundo Publishing Co.

JERIN, R.A., & MORIARTY, L.J. (1998). *Victims of crime.* Chicago: Nelson-Hall Publishers.

JERIN, R.A., MORIARTY, L.J., & GIBSON, M.A. (1995). Victim service or self-service: An analysis of prosecution based victim-witness assistance programs and providers. *Criminal Justice Policy Review, 7,* 142–154.

KALAHER, K. (1997). The proposed victim's rights amendment: Taking a bite out of crime or a dog with no teeth? *Seton Hall Legislative Journal, 22* (1), 317–355.

KARMEN, A. (1998). *Crime victims: An introduction to victimology (3rd ed.).* Belmont, CA: Wadsworth.

MATTHEWS, W.A. (1998). Proposed victims' rights amendment: Ethical considerations for the prudent prosecutor. *Georgetown Journal of Legal Ethics, 11,* 735–754.

MARION, N.E. (1995). The federal response to crime victims, 1960–1992. *Journal of Interpersonal Violence, 10*(4), 419–436.

NATIONAL ORGANIZATION FOR VICTIM ASSISTANCE. (1990). *Victim rights and services: A legislative directory 1988/1989.* Washington, DC: U.S. Department of Justice.

NATIONAL ORGANIZATION FOR VICTIM ASSISTANCE. (1993). A chronology of the victim's rights movement. *NOVA Newsletter, 16*(2).

NATIONAL ORGANIZATION FOR VICTIM ASSISTANCE. (1995). Victims rights week, 1995. *NOVA Newsletter, 17*(4).

NATIONAL ORGANIZATION FOR VICTIM ASSISTANCE. (1998). Rights amendment sent to senate. *NOVA Newsletter, 18* (4).

NEFF, C. (1997). Megan's law: Deterrence, remediation, or vengeance? *Maryland Bar Journal, 30* (5), 15–20.

PRESIDENT'S TASK FORCE ON VICTIMS OF CRIME. (1982). *Final report.* Washington, DC: U.S. Government Printing Office.

SABATUN, I.J., & EDWARDS, L.P. (1995). *Child abuse and the legal system.* Chicago: Nelson-Hall.

SCHLESINGER, S.R. (1984). *Victim/witness legislation: An overview.* Washington, DC: U.S. Department of Justice.

U.S. DEPARTMENT OF JUSTICE. (1998). *New directions from the field: Victims' rights and services for the 21st century.* Washington, DC: U.S. Government Printing Office.

WALLACE, H. (1998). *Victimology: Legal, psychological, and social perspectives.* Needham Heights, MA: Allyn and Bacon.

CASES AND STATUTES

Cases

- *Mapp v. Ohio*, 367 U.S. 641 (1961)
- *Gideon v. Wainwright*, 372 U.S. 335 (1963)
- *Miranda v. Arizona*, 384 U.S. 436 (1966)
- *Escobedo v. Illinois*, 378 U.S. 478 (1964)
- *In re Gault*, 387 U.S. 1 (1967)
- *Thurman v. Torrington*, 595 F. Supp. 1521 (Conn.1984)
- *Payne v. Tennessee*, 501 U.S. 808 (1991)
- *Michigan v. Lucas*, 500 U.S. 145 (1991)
- *United States v. Morrison* No. 99.5 (2000)

Statutes

- Victim Witness Protection Act of 1982 (18 U.S.C. 1512)
- The Victims of Crime Act of 1984 (18 U.S.C. 3013)
- Child Abuse Victims' Rights Act of 1986 (Title 18, Chapter 223, Sec. 3509)
- The Crime Control Act of 1990 (18 U.S.C. 225)
- Victims of Child Abuse Act of 1990 (18 U.S.C. 403)
- The Victims' Rights and Restitution Act of 1990 (42 U.S.C. 10606)
- Violent Crime Control and Law Enforcement Act of 1994 (18 U.S.C. 21)
- Violence Against Women Act of 1994 (Title 42, Chapter 136)
- Interstate Stalking Punishment Act of 1996 (Title 18, Chapter 110A, Sec. 2261A)
- Antiterrorism and Effective Death Penalty Act of 1996 (8 U.S.C. 1189)
- Mandatory Victims Restitution Act of 1996 (18 U.S.C. 3613A)
- Federal Drunk Driving Prevention Act (Title 23, Chapter 4, Sec. 408)
- Victim Rights Clarification Act of 1997 (Title 42, Chapter 112, Sec. 10606)

8

State and Federal Victim Resources and Services

Laura J. Moriarty

Robyn Diehl

INTRODUCTION

The International Association of Chiefs of Police (IACP) advocates police depart-ments following a model approach to victim service delivery (U.S. Department of Justice, 1997). The approaches suggested include either police departments developing law enforcement victim assistant (LEVA) units or establishing protocols that interface the police with service agencies.

LEVA units are organized specifically to address victims' needs. Dedicated personnel work directly with either all types of victims or victims of specific crimes, such as domestic violence victims. When police departments do not have specific LEVA units, often there are protocols in place that interface the police with service providers who are responsive to the needs of the victims. Departments often work with community service providers and other government agencies to provide a quick response to victims' needs. The provider often is not a police officer but someone who knows a great deal about the services available on the local and community level. The police officer plays a significant coordination role by making sure both parties are introduced.

The purpose of this chapter is to describe state and federal victim resources and services. Whether a police department has a separate LEVA unit or it relies on a proto-col where the department interfaces with social service providers in the community, it is important for all police officers to know what resources are available on the state and federal level.

WHY SHOULD THE POLICE BE CONCERNED ABOUT VICTIM NEEDS?

Victim satisfaction surveys have indicated victims are more supportive of the police, in particular, and the criminal justice system, in general, when the first contact with the system is positive, productive, and helpful. For many, the first contact with the criminal justice system is the police, in particular, the dispatcher. Therefore, the entire criminal justice system is viewed in a more favorable light if dispatchers and police officers are trained to address, minimally, the needs of victims. The dispatcher can help the victim by stating how long it will take for an officer to arrive, providing some general instructions as to what the victim should do while waiting for the police to arrive, and offering a number for the victim to call if the police do not arrive in a timely fashion. Having the dispatcher take a few extra moments to reassure the victim goes a long way to help the victim feel important, and cared for.

WHAT SIMPLE THINGS CAN BE DONE TO ADDRESS VICTIM NEEDS?

After the initial contact with the dispatcher, the police can do some very simple things to address victims' needs. Manilk and Stein (1992) summarized what crisis counselors do to help victims. They advocate police officers doing the same things.

Safety and Security

Many victims often report that no one at the crime scene ever takes the time to ask, "Are you all right?" Manilk and Stein believe this simple question will allow the officer an opportunity to display concern for the victim while encouraging the victim to talk, and to determine the victim's level of fear. If a person is frightened, the police officer may state, "You're safe now." Or, the officer may offer the advice of having the victim call a family member or friend to provide comfort, security, and relief.

Ventilation and Validation

Manilk and Stein (1992) support police officers taking the time to let the victim ventilate regarding the crime. Often victims have mixed emotional responses to victimization; they may cry or shake, become angry, or become numb. Whatever the reaction, Manilk and Stein agree that a short ventilation period is appropriate where the police officer expresses concern for the victim, and allows the victim to talk about the incident while expressing how he or she feels about it. It is also important for police officers to let the victim know that it is normal to feel scared, confused, or angry. This "validation" technique is the best and fastest way to calm a victim (Manilk & Stein, 1992).

Prediction and Preparation

Victims often describe their immediate feelings after a crime has occurred as having little or no control over their lives. Crisis counselors have suggested that police officers can help victims regain the sense of control over their lives by asking simple questions that are easily answered, such as, "May we sit over here?" or "Would you like a blanket?" Counselors maintain these small decisions help the victim regain a sense of control.

Counselors suggest that a victim will often focus on why this particular event happened to him or her. They suggest that police officers avoid blaming the victim and focus the blame on the offender: "He had no right to take your wallet." Focusing on the offender is another way to restore the victim's sense of control.

Another way is to provide a clear and concise explanation of what will happen next. Let the person know a detective will be coming by shortly, and his or her name. Conversely, if nothing is going to happen for some time, let the victim know that as well. Be honest.

WHAT IS THE TYPICAL RESPONSE OF A CRIME VICTIM?

Although not all crime victims will respond to crime in the same manner, there are some common elements found among most crime victims. Bard and Symonds (as cited in Manilk & Stein, 1992) were among the first to describe the crime victim's crisis reactions as occurring in stages. The most important stage for police officers is the first stage lasting from a few hours to several days. The victims are typically in shock. They are disorientated. They have a sense of helplessness, fear, and disbelief (Manilk & Stein, 1992). Knowing that this is the most common and first response to the crisis of being a crime victim, the police officer is reminded of the previous section where three simple steps were suggested to help meet the needs of the crime victim.

WHAT DO VICTIMS EXPECT FROM THE POLICE?

It is apparent that crime victims want a caring and compassionate response to their victimization. Following the suggestions listed previously will add greatly to victims feeling that the police really want to do something about crime and are interested in their plight.

Victims want accurate information from the police, even if the information is not pleasant: "No, you probably will never recover your stolen property." Victims would rather hear accurate, truthful statements than false ones that build false hopes and are later destroyed.

Additionally, victims expect the police to be able to assist them by being informed of victim services available in the community, on the state level, and even the national (federal) level. In Chapter 7, Jerin provided an overview of the legislative enactments that resulted in certain rights afforded to all victims. In this chapter, we present national organizations that assist victims. The primary services provided by these organizations are listed as well. Next, we highlight one specific state, North Carolina, listing and discussing the statewide resources available to victims. It is imperative for all police officers to be familiar with such resources.

National Resources

This section provides an overview of national resources available to help victims of crime (Young, 1993).

Murder or Homicide.   Survivors, including family members, should be counseled to seek support from mental health professionals and/or self-help groups for their emotional well-being. As discussed in Chapter 2, secondary victims (in this case, those whom the deceased is survived by) also need care. There are three national agencies:

1. **Parents of Murdered Children**

   Parents of Murdered Children and other survivors of homicide victims (POMC) has chapters in most states. To find the chapter in your state, write or call:

   > POMC National Headquarters
   > 1739 Bella Vista
   > Cincinnati, OH 45227
   > (513) 731-LOVE

2. **The Compassionate Friends**

   The Compassionate Friends help parents who have lost their children by any cause of death including homicide.

   > The Compassionate Friends National Headquarters
   > P.O. Box 1347
   > Oak Brook, IL 60521
   > (312) 323-5010

3. **They Help Each Other Spiritually (THEOS)**

   THEOS is an organization that helps spouses deal with the death of a spouse regardless of the cause of death.

   > THEOS Foundation
   > 410 Penn Mills Mall
   > Pittsburgh, PA 15235
   > (412) 243-4299

Vehicular Homicide.   Mothers Against Drunk Driving (MADD) is perhaps the best known organization to assist with victim advocacy related to vehicular homicide caused by drunk driving. There are many local chapters.

> MADD National Headquarters
> 511 John Carpenter Freeway
> Irving, TX 75345
> 1-800-GET-MADD

Other National Resources.   Figure 8-1 lists other national resources. Because victims experience many and diverse reactions to criminal victimization, these resources are listed in order for the victims to decide if the services provided by the agency are needed in their particular case. Rather than just giving the victim one number, the police office should reproduce Figure 8-1 as a small card, and give all the resources listed to the victim.

Administration on Aging Eldercare Locator	1-800-677-1116
American Indian Programs and Services	1-202-208-3711
Childhelp USA/Forrester National Child Abuse Hotline	1-800-4A-CHILD
The Compassionate Friends National Headquarters	1-312-323-5010
Deaf, Hard of Hearing and Speech Disabled Assistance	1-800-877-8339
Domestic Violence Hotline	1-800-799-7233
Family Violence Prevention Fund/ Health Resource Center	1-800-313-1310
Health and Human Services	http://www.dhhs.gov
MADD National Headquarters	1-800-GET-MADD
Maternal and Child Health Information	1-703-356-1964
Medicare—General Information	1-800-638-6833
Mental Health Information	1-800-789-2647
National CASA (Court Appointed Special Advocates)	1-800-628-3233
National Center for Missing and Exploited Children	1-800-843-5678
National Child Abuse and Neglect	1-800-422-4453
National Crime Prevention Center	1-800-937-7383
National Resource Center on Child Sexual Abuse	1-800-543-7006
National Runaway Switchboard	1-800-621-4000
Office for Victims of Crime Resource Center	1-800-627-6872
POMC National Headquarters	1-513-731-LOVE
Rape, Abuse & Incest National Network	1-800-656-4673
Resource Center on Child Protection and Custody	1-800-527-3223
Sexually Transmitted Diseases Hotline	1-800-227-8922
They Help Each Other Spiritually (THEOS)	1-412-243-4299
Women's Health Information	1-800-232-3299

*Compiled from various Internet sources including the NOVA homepage (www.nova.org).

FIGURE 8-1   National resources to assist victims of crime.*

Statewide Services

All states provide services to victims. Most often the services are provided by victim and witness assistance agencies, often housed in the district attorney's office or in law enforcement units.

Here we examine one state, North Carolina, and list all the statewide victim services that are available to help. Similar services are available in all states. The first point of contact should always be the victim advocate. All police officers should know the victim advocate and refer victims to that person.

In North Carolina, the victim advocate is part of a wider network called the North Carolina Victim Assistance Network. This organization provides statewide information and referral for victims of crime. Other agencies providing assistance to victims follow.

Each agency is explained in this section, followed by the services provided. This information is found in the *Directory of Victim Services, Emergency Resources, and Related Criminal Justice Agencies in North Carolina* North Carolina Victim Assistance Network, (1993).

Office of the Attorney General.   The attorney general is responsible for two main areas: legal services and law enforcement. The attorney general consults with and advises judges, district attorneys, magistrates, and municipal and county attorneys.

Citizens' Rights Division.   The Citizens' Rights Division of the Office of Attorney General ensures that the attorney general's office advocates and protects the interest of the public.

Office of Citizen Services: Ombudsman Program, CARE-LINE Information, and Referral Service.   The Department of Human Resources Statewide provides a toll-free information and referral service (CARE-LINE).

Department of Crime Control and Public Safety, Division of Victim and Justice Services.   Victim and Justice Services is responsible for overseeing the Victim Compensation Commission, which reimburses citizens who suffer a financial loss as a result of a crime.

Governor's Crime Commission, Division of Crime Prevention.   The Crime Prevention Division serves as the state's central agency focusing on the prevention of crime. The goal of the division is to empower and involve citizens in the fight against crime by establishing school, community, county, and statewide crime prevention programs.

Administrative Office of the Courts.   The function of Administrative Office of the Courts can be grouped into several major categories including fiscal management, personnel direction, information and statistical services, juvenile services, guardian ad litem services, trial court scheduling/management services, research, planning, and administrative services.

Conference of District Attorneys.   Under the Administrative Office of the Courts, the Conference of District Attorneys provides training opportunities for the state's district attorneys and staff. The conference also monitors the Victim Witness Assistance programs statewide.

Guardian Ad Litem Division Program. This program is under the Administrative Office of the Courts. Guardians ad litem and attorney advocates are appointed by juvenile/district court to represent the interest of children who are allegedly being abused or neglected. Guardians ad litem and the attorneys represent the children for the duration of the court proceedings.

North Carolina Department of Administration Disability Hotline. This hotline provides information about the application process for disability pay and pending cases.

MADD North Carolina State Office. The mission of MADD is to stop drunk driving and support the victims of this violent crime. The state office offers community education, information and referral, and victim advocacy.

North Carolina Coalition Against Domestic Violence. The coalition provides community education and technical assistance to domestic violence programs statewide.

North Carolina Coalition Against Sexual Assault. The coalition provides community education and technical assistance to sexual assault programs statewide.

Prevent Child Abuse. This service provides statewide information and referral for child abuse issues as well as a resource library.

Division of Social Services. The Division of Social Services provides numerous emergency services to the residents of North Carolina.

North Carolina Bar Association's Lawyer Referral Service. This lawyer referral service is free.

North Carolina Association of Residential Child Care and Family Services. This is a statewide association of agencies that provide residential child care.

Local Services

Other services are available on the local level. As an illustration, we selected one county in North Carolina to detail its victims' services. Police officers should familiarize themselves with the local (community) resources available to victims of crimes. Again, the first place to begin is with the victim advocate but police offices can also consult the local phone book to determine if similar services are available in their community.

Each agency is explained in this section, followed by the services provided. This information is found in the *Directory of Victim Services, Emergency Resources, and Related Criminal Justice Agencies in North Carolina* North Carolina Victim Assistance Network, (1993).

Family Services Center.   The center offers domestic violence/sexual assault services, a general counseling center, family mending for families affected by child sexual assault, and support groups. The center serves victims of child sexual abuse and adult survivors of child sexual assault.

Guardian Ad Litem.   In addition, guardians ad litem work with the justice community agencies to locate and develop resources for children.

Helpmate, Inc.   Helpmate provides a 24-hour crisis line, community education and training, individual and group support counseling, court advocacy/accompaniment, emergency transportation, a children's program, and a shelter with a negotiable fee.

Rape Crisis Center, Inc.   This organization provides a 24-hour crisis line, support groups, court advocacy, information and referral, victim assistance for medical expenses, community education and training, and children's programs.

Thoms Rehabilitation Center.   The center provides physical rehabilitation services for physically challenged individuals.

Western North Carolina Regional Child Abuse Center.   The center provides child medical exams for victims of alleged sexual abuse, prevention education, information and referral, parenting information, and a resource library.

Asheville Head Injury Families.   This support group is for individuals who have experienced head injuries.

Compassionate Friends, Mountain Area Chapter.   Compassionate Friends is a grief group for parents and siblings who have lost a child/sibling. It is a national association with local chapters.

First Call for Help.   This countywide information and referral service provides an on-call answering service for adult/child protective services, mental health, rape crisis, domestic violence, and grief services.

MADD Buncombe County Chapter.   Services include anti–drinking and driving campaigns, victim advocacy, victim impact panel, court watch, and a speakers bureau. MADD is a national association with local chapters.

Mountain Area Hospice.   The hospice offers bereavement support services.

Parents of Murdered Children: Western North Carolina Chapter.   This self-help organization is designed solely to offer emotional support and information about surviving the loss of a child to murder. It is a national association with local chapters.

Asheville Area Chapter American Red Cross.   The Red Cross provides disaster relief services, blood supply services, health and safety training, and services to military personnel and their families. It is a national organization with local chapters.

Asheville-Buncombe Community Christian Ministry.   The ministry provides emergency food, emergency assistance programs, free/low-cost fuel, and emergency shelter.

Buncombe County Caring for Children.   This organization provides emergency shelter for children.

Buncombe County Salvation Army.   The Salvation Army provides emergency food, emergency assistance programs, free/low-cost fuel, and emergency shelter. It is a national organization with local chapters.

Hospitality Home of Asheville.   Hospitality Home is an emergency shelter.

Western Carolina Rescue Mission.   The rescue mission is an emergency shelter.

Buncombe County Department of Social Services.   The Department of Social Services provides counseling services, adoption services, adult day care, adult in-home services, child and adult protective services, financial and medical assistance to low-income families, emergency assistance with utility bills and rent, food distribution, information about food stamps, custody studies, foster care, information about Medicaid, parenting skills services, and information and referral. It is a federal organization with local departments.

Blue Ridge Mental Health Center.   The center provides 24-hour emergency services; developmental disability services; counseling for individuals, couples, and families; substance abuse services; and community support programs for mentally ill adults.

Asheville District Office Social Security Administration.   The Social Security Administration is responsible for social security disability, retirement, and survivors' benefits.

North Carolina Division of Vocational Rehabilitation Services.   Services are provided to persons with physical, mental, or emotional impairments that result in a substantial impediment to employment. Services include disability and vocational assessment, employment training, counseling, and placement.

CONCLUSION

To this point in the text, the focus has been to provide information to police officers to help them do their jobs more efficiently and effectively from the perspective or viewpoint of the victim. In this chapter, we focused on victim resources and services available at the local, state, and federal levels to assist victims of crime. Whether a police department has developed a unit specifically to address victims' needs or it has established protocols where community services are interfaced with the department allowing officers to refer victims to certain individuals, it is still imperative that police officers know more about services and resources available to victims. Often victims need immediate care and direction on how to handle the situation.

The information presented in this chapter will allow the police officer to know, in a general sense, what services and resources are available in his or her community and on the state level to assist victims. Moreover, the national organizations listed have local and/or statewide chapters. Developing a brochure or card, similar to those used to ensure that victims know their rights (see Chapter 7, Figure 7-1), is beneficial for all parties. The police officers have something to refer to when they need to be refreshed, and victims feel like the police care because they listed resources and services that will in all likelihood provide additional help and assistance.

Figure 8-1 can easily be reproduced as a small card providing officers with national organizations that assist victims. Additionally, police officers should take a few minutes to peruse the local telephone book to determine if the national organizations have local chapters. Recording the local numbers on the back of the card is one simple strategy for increasing the victim's awareness of services available in the local community.

REFERENCES

MANILK, B.A., & STEIN, J. (1992). *Police victim assistance: Concepts and issues paper.* Alexandria, VA: IACP National Law Enforcement Policy Center.

NORTH CAROLINA VICTIM ASSISTANCE NETWORK. (1993). *Directory of victim services, emergency resources, and related criminal justice agencies in North Carolina.* Raleigh, NC: Author.

U.S. DEPARTMENT OF JUSTICE (1997). *Law enforcement's promising practices in the treatment of crime victims: A report to promote the development and expansion of victim-oriented policies and practices in law enforcement agencies.* Washington, DC: Author.

YOUNG, M.A. (1993). *Victim assistance: Frontiers and fundamentals.* Washington, DC: National Organization for Victim Assistance.

9

Campus Policing and Victim Services

Max L. Bromley

Bonnie S. Fisher

INTRODUCTION

College life has been part of the fabric of American society since colonial times when attending college was the province of a privileged minority. This elite characterization of higher education, however, changed at the end of World War II when thousands of military personnel returned to civilian life. A significant proportion of these veterans took advantage of the financial assistance included in the Servicemen's Readjustment Act—known as the GI Bill (Public Law 78-346)—to obtain a college degree. As a result, the number of two-year and four-year post-secondary institutions increased dramatically and student enrollment rose significantly.

Subsequently, the "baby boom" generation matured to college age during the 1960s and 1970s. Boomers enrolled in colleges and universities at record numbers with help from the federal government in the form of low-interest student loans and work-study programs, and other federal legislation such as the Civil Rights Act of 1964 (Public Law 88-452). Once again, the composition of colleges and universities changed—total student enrollment increased, as did enrollments for female and minority students. Changes in all three of these enrollment characteristics are evident in the composition of college students in the 1990s. For example, the total fall enrollment for 1995 was 14.3 million students, up 16% from 1985. During this time period, the number of females enrolled in undergraduate studies rose by 23% to nearly 8 million and the number of females enrolled in graduate schools increased 38% to nearly 1 million. Further, the proportion of minorities also increased significantly from 16% in 1976 to 25% in 1995 (U.S. Department of Health, Education, and Welfare, 1997, tables 176, 187, 188, 189, and 206).[1] Likewise, faculty and staff increased to accommodate the influx of new students. In the fall of 1993, approximately 2.6 million people were employed in colleges and universities (U.S. Department of Health, Education, and Welfare, 1997, table 223).

College students are enrolled in a variety of different types of post-secondary institutions from "traditional" four-year residential schools to schools that are less than two years (e.g., cosmetology, vocational-technical schools, business and computer programming schools, and health careers schools). Although most institutions are small in terms of student enrollment, most students attend large institutions. The largest proportion of students attend public four-year and two-year schools (40% and 36% of students, respectively) (Lewis & Farris, 1997).

One of the unanticipated consequences of the tremendous growth in higher education has been a concern about the extent of on-campus property crimes, violent crimes, and alcohol and drug-related violations. Smith (1989, p. 10) observed, "As the sizes of institutions grew and the students came to more closely represent a cross-section of the social and economic classes of the nation, the incidents of campus crime likewise increased." More to the point, Carrington (1991, p. 50) summarized the campus crime issue when he remarked ". . . put to rest the long-cherished notion that colleges and universities are somehow cloistered enclaves—sanctuaries far removed from the threat of crime that haunts the rest of us."

Observations like these have led some educators and researchers to abandon their traditional "ivory tower" image of campuses in lieu of another image of campuses—"hot spots" for criminal activity. Some campus experts have dubbed campuses "dangerous places" (Smith & Fossey, 1995) and "armed camps" (Mathews, 1993) where firearms are a growing menace that have created deadly effects when mixed with alcohol and drugs (see also Miller, Hemenway, & Wechsler, 1999; Whitaker & Pollard, 1993).

In response to the changing image of college and university campuses, the role and functions of campus police have evolved over the last 100 years to meet the security, safety, and victim service needs of the campus community—faculty, staff, students, and visitors. To gain a broader understanding of the roles and functions of campus policing and victim services, we first trace the evolution of the campus police department from night watchmen–security guards to law enforcers–service providers. Secondly, we discuss the rise of the awareness to campus victimizations and victim advocacy as national-level and state-level policy issues and their effects on campus police and victim services. Here we also offer an overview of the results from the renewed research interest in campus victimization. Thirdly, we examine how colleges and universities and campus police departments have responded to the needs and mandates for victim services and include two case study examples. Finally, in closing, we discuss challenges encountered when dealing with victims of campus crimes and suggest several recommended steps to be taken by campus police and other service providers. In doing so, we also identify issues that need to be more fully examined with regard to providing quality service to victims of campus crime. Unless we note otherwise, throughout our chapter, we will refer to the two types of schools (i.e., public four-year and two-year institutions) that the majority of students attend.

---

[1]Minorities include Asian and Pacific Islander, Hispanic, and African American students. These percentages exclude foreign students who are enrolled in U.S. colleges and universities.

THE EVOLUTION OF CAMPUS POLICE: FROM NIGHT WATCHMEN–SECURITY GUARDS TO LAW ENFORCERS–SERVICE PROVIDERS

The evolution of campus police from night watchmen–security guards to law enforcers–service providers has been largely influenced by three factors: (1) increased student enrollment, (2) political events external to the campus, and (3) the changing nature of crime on campus. First, as we previously noted, student enrollment has steadily increased over time and with these increases, a number of safety, security, and victim service issues arose. Second, campus administrators and police were forced to react to political events that eventually transformed campus police departments. Finally, over time, the extent and nature of crime on campuses has changed from being considered youthful pranks to legally defined as acts of violence. We now turn to discussing how each of these factors influenced the evolution of campus police and victim services.

The Early Years of Campus Police

The origin of the modern campus police department was in 1894. Yale University administrators hired two officers to police the campus in response to frequent conflicts between New Haven residents and Yale students. Yale's campus police "department" was the exception during the early 1900s; most colleges and universities depended on local police to handle any criminal incidents or campus disruptions, and student misbehavior typically was handled by the dean of students' office (Nichols, 1997). In the early part of the 20th century, campus security employed a large number of retired police or military officers who performed primarily night watchman–security guard activities. Their main responsibilities were the protection of university and student property and the monitoring of student conduct (Esposito & Stormer, 1989).

With the repeal of Prohibition in the early 1930s, the night watchman–security guard approach evolved into an enforcement approach—the enforcement of rules and regulations that governed student conduct (Nichols, 1997). During the 1940s and 1950s, campuses were still characterized as relatively free from serious crime although problems created by the increased student enrollment were emerging. The frequency by which campus police had to respond to parking problems and student alcohol-related infractions as well as maintain building security signaled to campus administrators that there was a pressing need for an organized, professional police presence on campus (Nichols, 1997; Powell, 1981).

The Birth of the Modern Campus Police Department

The 1960s and 1970s.   During these two decades, campus administrators realized that campus security had outgrown its night watchman–security guard origins. National and international events such as the Civil Rights movement in the 1960s and the anti–Vietnam War movement in the 1970s created a volatile environment on many campuses. Given the dominance of the night watchman–security guard approach, most campus police departments were ill-prepared for the massive demonstrations, property destruction, and personal violence that occurred on numerous campuses during this time (Powell, Pander, & Nielsen,

1994). Utilizing local police or the National Guard to deal with these problems on campus often led to disastrous results, including the killing of four students at Kent State University (Peak, 1995).

It was during the late 1960s and early 1970s that a major shift in the evolution of campus policing occurred. Sloan (1992b, p. 87) observed that "the modern campus police department" was formed "as campus unrest grew and the specter of urban police on college campuses loomed, college administrators confronted a dilemma." Administrators recognized how inadequate campus security was in addressing the safety and security needs of the campus community but were hesitant to overrely on local police given the all too known problems with the local police and campus relationship. They had come to realize the need for a professionally trained, well-organized campus police department that could serve the safety, security, and service needs of the campus community.

During this time, many campus administrators upgraded campus security by hiring more educated and fully trained police personnel. With this new cadre of officers, schools began to rely less on local police for law enforcement responsibilities on campus and turn more to their own police personnel. While these changes in police departments were occurring on campus, many states enacted laws that granted full police authority to officers employed at public institutions of higher education (Bromley, 1996). Other states granted this authority through local deputization (Gelber, 1972 as cited by Bromley, 1996). With these internal and external changes, the role of the campus police officer had moved from a night watchman–security guard perspective to an orientation that more closely resembled a professional police officer model (Sloan, 1992b; Webb, 1975). By the mid-1970s, student enrollment continued to grow as campuses continued to experience an increase in the incidences of crime, in particular more violent crimes such as murder and sexual violence (Smith, 1995). Some experts argued that this upward trend in campus violence continued into the 1980s, whereas others argued that campus violence decreased (Fernandez & Lizotte, 1995; Nichols, 1986).

The 1980s and 1990s.   During the late 1980s and into the 1990s, forces external to the campus began to focus greater attention on the campus crime "problem," and heightened concern about campus victims came to the forefront of the policy agenda of several interested parties. In response to the efforts of these parties, today most of the nearly 6,000 institutions of higher education have their own campus public safety or law enforcement departments that are responsible for providing a broad range of police, facilities and special events security, parking management, crime prevention, and victim services (Reaves & Goldberg, 1996).

Two studies provide insight into the current state of the modern campus police department. First, the Department of Education published a study of campus crime and security at post-secondary institutions by the National Center for Education Statistics (NCES) (Lewis & Farris, 1997). Using a nationally representative sample of 1,218 post-secondary, federal Title IV eligible institutions (e.g, four-year, two-year, and less-than-two-year; public and private; for-profit and private nonprofit), they reported that 46% of these schools employed sworn officers (i.e., officers with full arrest power), 34% employed security officers or guards, 24% had contracted for security, 8% used the city or state police when called (e.g., through the use of 911 or other local emergency numbers), and 15% provided security through other means. Their study further reported that much variation exists in the types of public safety employees used by different types of institutions. For example, 80% of public four-year institutions used sworn officers as compared to 6% of private two-year institutions. Even among four-year institutions, there is variation as only 17% of the private four-year institutions employ sworn officers.

Second, a U.S. Department of Justice study conducted by Reaves and Goldberg at the Bureau of Justice Statistics (BJS) of 680 campus law enforcement agencies located at four-year schools in the United States with 2,500 or more students reported that 93% of all campus law enforcement departments serving public institutions used sworn officers, including all schools with 25,000 or more students (Reaves & Goldberg, 1996, p. iii).

The majority of campus police departments perform many of the same operational and administrative functions as their municipal police counterparts, including having arrest authority and being armed. The BJS study also reported that 75% of the schools in their sample had officers with arrest authority and 64% had armed officers (Reaves & Goldberg, 1996, table 1). Many campus police departments have followed their municipal counterparts and have adopted community policing, foot patrol, and bike patrol (Lanier, 1995; Reaves & Goldberg, 1996). An NCES study reported that a majority of law enforcement/security personnel at four-year public (93%), private four-year (71%), and two-year public post-secondary institutions used foot or bicycle patrols (Lewis & Farris, 1997).

Despite their similarities in functions, experts have observed that campus police emphasize service-related duties, are more crime prevention oriented, and are less focused on enforcement duties than their municipal law enforcement counterparts (Lanier, 1995). The BJS study supports this observation: 78% of full-time sworn personnel were regularly assigned to responding to calls for service (Reaves & Goldberg, 1996, p. 5).

The Community Policing Movement within Campus Police Departments.   Although there are varying definitions of community-oriented policing, there are three generally accepted characteristics associated with this philosophy. These characteristics are as follows.

1. The creation of and reliance on effective partnerships with the community and other public and private sector sources.
2. The application of problem-solving strategies and tactics.
3. The transformation of the organizational structure and culture of policing to support this philosophical shift (Bureau of Justice Assistance, 1997).

Recognizing the benefits of working with the campus community to address crime problems and victim service needs, many campus departments, like municipal police departments, have implemented strategies based upon the principles of community-oriented policing (COP). In these departments, the role of police is broadened beyond the narrow focus of law enforcement as police officers seek to work closely with campus community members and off-campus professionals on a variety of quality-of-life issues. Several campus police departments have been practicing COP for years and have demonstrated the effectiveness of this philosophy. The University of Cincinnati, the University of Washington, Michigan State University, Virginia Commonwealth University, and the University of South

Florida are examples of higher education institutions whose police departments are actively involved in community-oriented policing. These universities have tailored their COP efforts, based upon community input, to meet the safety, security, and victims' needs of their respective campus community members (Benson, 1993; Johnson, 1995; Lanier, 1995). The emphasis on determining the safety, security, and victim needs of the campus community, and then cooperatively working with campus community members, including campus administrators, to meet those needs is central to any COP effort.

Some campus police departments assign police officers to specific geographic locations for long periods of time to develop cooperative problem-solving strategies with students and staff members. Michigan State University, the University of Alabama at Birmingham, and the University of South Florida, for example, have student housing campus police officers within the student residence hall areas to enhance the interactions among students, the housing staff, and the respective campus police department. The campus police officers assigned there are in position to facilitate the needs of the students before a situation becomes a criminal one, to establish cooperative working relationships, to encourage students to report crime, and to offer a more personalized approach in working with the needs of victims or those who know victims. These officers can also work closely with students to develop strategies to reduce the likelihood of them becoming a victim and conduct seminars for students on various subjects ranging from date rape, to holiday safety, to personal awareness (see 1999).

An officer assigned to a student residence hall area on a routine basis can be an important link in dealing with victim-related issues. For example, on the campus of one of the authors, a police officer assigned to a residence hall area recently had to inform the students that three of their fellow residents had been tragically killed in a traffic accident near the campus. Based upon his personal knowledge of the residence hall population, this officer was able to personalize this sad message in a way that would have been impossible for a police officer who was a total stranger. Practicing community-oriented policing on a campus creates a much better opportunity to provide quality victim services as opposed to a traditional law enforcement focus.

Although the municipal community policing movement influenced the operations of campus policing, there are other external forces to the campus that have specifically targeted the safety, security, and victim needs of college and university campuses. The media, the courts, grassroots efforts, federal and state legislative bodies, and researchers have played a significant part in evolution of campus police and their role and functions in providing different types of victim services. We now turn our discussion to the efforts of these parties and the impact that they have had on campus policing and victim services.

THE RISE OF AWARENESS OF CRIME AND VICTIM ADVOCACY ON CAMPUSES AND ITS EFFECTS ON CAMPUS POLICE AND VICTIM SERVICES

During the last 15 years, various parties interested in crime on college and university campuses have been successful in capturing the attention of policy makers at the state and federal levels. Their actions have resulted in heightened awareness of campus crime and have led to the passage of federal-level and state-level legislation, the emergence of civil litigation, and a renewed research interest in the extent and nature of campus crime and the roles and functions of campus police in providing victim services.

The Media's Coverage of Campus Crime Incidents

Much of the public's awareness of campus crime has come largely from intensive media coverage of sensational crimes, usually involving students and sometimes faculty members as victims (Brantingham, Brantingham, & Seagrave, 1995; Fisher & Sloan, 1995). Though serious crimes such as murder and rape are relatively infrequent incidents on campus compared to property crimes, they draw considerable attention from both print and electronic media when they do occur (Bromley, 1998; ww.soconline.org, 1999).

Over the last decade, several nationally respected newspapers including *The New York Times* and *USA Today* (Castelli, 1990; Kalette, 1990; Mathews, 1993; Ordovensky, 1990) have printed special features on the extent and nature of campus crime. One of the well-respected higher education publications, *The Chronicle of Higher Education,* routinely highlights various issues related to campus victimizations including legislative action and court decisions. Since 1993, there has been an annual section in *The Chronicle of Higher Education* devoted to campus crime statistics and arrests for drug-law, alcohol-law, and weapons-law violations (Lederman, 1993, February 1994, March 1994; Lively, April 1996, May 1996, 1997; Nicklin, 1999).

The media have certainly played a major role over the last decade in describing serious, violent crimes that have occurred on college campuses. Although it has been important to raise the general public awareness, some have suggested that the focus on sensational, infrequently occurring violent crimes has given the general public the impression that the typical campus is more dangerous than it really is (Brantingham et al., 1995). This in turn has led to concerns by parents, students, campus administrators, and policy makers about the safety and security needs of the campus and the types of victim services needed for campus victims (Fisher & Sloan, 1995).

Civil Court Decisions

The civil courts have also played a major role, not only in raising the level of public awareness about campus crime victimization, but also in ruling that colleges and universities can be held liable for "foreseeable" criminal acts that happen on campus. These decisions have forced college administrators to acknowledge that serious crimes can occur even in the relatively safe confines of a college campus and to address the causes of these crimes.

Since the late 1970s, campus crime victims and their families have filed a considerable number of civil lawsuits against colleges and universities, claiming that the latter were negligent with respect to providing a safe and secure campus. According to Smith (1995, p. 26), campus crime victims and their families can sue colleges and universities under one of the following duty categories: (1) a duty to warn about known risks, (2) a duty to provide adequate security protection, (3) a duty to screen other students and employees from dangers, and (4) a duty to control student conduct.

Early landmark cases such as *Duarte v. State of California* (1979), *Peterson v. San Francisco Community College* (1984), *Miller v. State of New York* (1984), and *Mullins v. Pine Manor College* (1983) are examples of cases won by campus crime victims or their families. They are precedent-setting decisions and underscore the premise that institutions of higher education do have duties with regard to taking reasonable steps to prevent "foreseeable" crime as well as to provide an adequate level of security (Smith & Fossey, 1995).

In the 1990s three additional civil lawsuits underscored the university's continuing role in providing adequate campus security. First, in *Johnson v. Washington* (1995), a first-year student was abducted and sexually assaulted on the campus of Washington State University. The court determined that the student was owed a duty of reasonable care as an invitee and tenant (the attack occurred near her dormitory). The university was required to exercise reasonable—not all possible—care once criminal danger in the area became foreseeable. Johnson also (re)affirmed two points: first, the attacker was not the sole proximate cause as a matter of law, and second, "the duty to protect against foreseeable criminal attack was not subject to discretionary governmental immunity."

Second, in *Nero v. Kansas State University* (1993), a student was sexually assaulted by another student in the basement television room of her dormitory. The attacker had been placed in the dormitory despite having been accused and charged with rape and sexual assault and despite the fact that he had previously been removed from a male/female residence and placed in an all-male dormitory. There were no warning signs given and no special safety precautions taken. The Nero court determined that a college must use reasonable care once a dangerous student is placed in residential housing.

Third, in *Gross v. Family Services Agency, Inc.* (1998), Gross, a 23-year-old graduate student, was attacked in the parking lot of an off-campus "practicum" location. The student was required to do a practicum and had selected this location from a list provided by her university. Here, a Florida intermediate appellate court ruled that a student participating in an off-campus internship is entitled to reasonable care from the sponsoring university with regard to foreseeable danger.

Thus, campus crime victims have certain legal protections afforded them that go beyond that which is generally provided to citizens in the general community. This factor is an important distinction and by extension requires campus police and other campus officials to be more proactive than municipal authorities in dealing with victimization issues.

The Clerys' Grassroots Movement

Perhaps no lawsuit against a college for failing to provide adequate security has received more notoriety or achieved more long-term results than that brought by the Clery family against Lehigh (Pennsylvania) University in 1986. Following the brutal murder-rape of their daughter on the campus of Lehigh University, the Clerys sued the university and won an out-of-court settlement against the institution.

The loss of their daughter ignited a far-reaching cause—to bring the problem of violent crime on campus to the attention of faculty, staff, and current and prospective students (U.S. Department of Justice, 1998). Recognizing the lack of security and victim services provided by most campuses, the Clerys founded Security On Campus, Inc. (SOC) in 1987. SOC is the only "national, non-profit organization geared specifically and exclusively to the prevention of college and university campus violence and other crimes". . . and "dedicated to assisting campus crime victims in the enforcement of their legal rights" (www.soconline.org, 1999). The Clery family has used SOC as their medium to influence campuses to change their security policies and develop victim services. They launched an aggressive campaign to heighten awareness of campus crime and to secure the passage of federal-level and state-level legislation that would require colleges and universities to report their crimes publicly, develop crime prevention strategies to reduce victimization, and provide services for campus crime victims.

The Clerys' grassroots movement eventually led to the passage of the first campus crime reporting statute by the Pennsylvania State Legislature in 1988. Fueled by this victory, their crusade moved to the federal level. In 1991, their efforts led to the passage of the *Student Right-to-Know and Campus Security Act* (Public Law 101-542), the first federal-level legislation to address the reporting of campus crime statistics (Fisher, 1995; Griffaton, 1995). Following these two pieces of legislation, over a dozen other states subsequently enacted campus crime reporting statutes (Fisher, 1995). Several state-level statutes (e.g., New York, Washington, Wisconsin, and California) also establish specific requirements for campuses with regard to providing services to campus crime victims. Not only are these federal-level and state-level laws important in raising the level of public awareness about campus crimes and safety on campus, but additionally these laws also require campus police and other college officials to play an important role in victim service development and delivery. Examples of campus police being involved in campus-specific promotion of victim-related commemorative observances include National Crime Victims' Rights Week, Sexual Assault Awareness Month, and National Drunk and Drugged Driving Awareness Week (National Criminal Justice Association, 1999).

The Passage of Federal-Level and State-Level Legislation

Federal-Level Legislation.   At the federal level, Congress passed the Student Right-to-Know and Campus Security Act (hereafter referred to as the Act) and President Bush signed it into law in 1990. The Act and its subsequent amendments mandated requirements for all higher education institutions, both public and private, that participate in federal Title IV financial aid programs to publicly disclose their respective campus crime statistics, security policy, and victim services. The Act states that schools must "prepare, publish, and distribute, through appropriate publications or mailings, to all current students and employees, and to any applicant for enrollment or employment upon request, an annual security report" containing campus security policies and campus crime statistics for that institution (see 20 U.S.C. section 1092(f)(1)) (see Sloan, Fisher, & Cullen, 1997). The original law required schools to publish statistics for murder, rape, robbery, aggravated assault, burglary, and motor vehicle theft. In addition, schools had to publish arrests for liquor-law, drug-law, and weapons-law violations. The law was amended by the Higher Education Act of 1992 to require schools to replace the rape statistics with data for forcible and nonforcible sex offenses.

The Act was the first federal mandate to require colleges and universities to publicly report their crime statistics. Prior to the passage of the Act there was no requirement for schools to report any crime to the Federal Bureau of Investigation (FBI) for inclusion in its Uniform Crime Report (UCR). Since 1972 when the FBI began including colleges and universities in the UCR, very few schools reported to them. In 1991, for example, only 12% of all colleges and universities in the United States reported their crime statistics to the UCR. As a result, little was actually known about what and how many crimes were known to campus police.

In 1991, Congress amended the Act to include the *Campus Sexual Assault Victims' Bill of Rights,* which requires colleges and universities to develop and publish as part of their annual security report policies regarding the awareness and prevention of sexual assaults and to afford basic rights to sexual assault victims (campussafety.org/cas/right.html, 1999). These rights include the following procedures: (1) the "accuser and accused must have the same opportunity to have others present (*sic* at the disciplinary hearing), (2) both parties shall be informed of the outcome of any disciplinary proceeding, (3) survivors shall be informed of their options to notify law enforcement, (4) survivors shall be notified of counseling services, and (5) survivors shall be notified of options for changing academic and living situations" (campussafety. org/CSA/rights.html, 1999). In addition to counseling services, students must also be informed of medical and legal assistance, if they become a victim (U.S. Department of Justice, 1998).

The Act was amended again in 1998 to include additional reporting obligations (e.g., manslaughter and arson), extensive campus security-related provisions, and the requirement of keeping a daily public crime log for the first time (some states already require a public log, see Fisher 1995; Griffaton, 1995) (Public Law 105-244). The 1998 amendments also officially changed the name of the Act to the *Jeanne Clery Disclosure of Campus Security Policy and Campus Crime Statistics Act* (hereafter referred to as the Clery Act) (20 U.S.C. section 1092).

In the fall of 1999, the U.S. Justice Department awarded $8.1 million to 21 colleges and universities to develop and implement and/or improve campus-based programs and services for victims of sexual assault, domestic violence, and stalking. The universities and colleges receiving these funds must train campus police about responding to sexual assault, domestic violence, and stalking, and establish a mandatory prevention and education program on violence against women for all incoming students. Many of these schools are developing comprehensive victim response systems that include the campus police working with faculty and student organizations, health and counseling services, and local law enforcement. According to Attorney General Janet Reno, these grants will ". . . enable schools to better respond to victims, provide needed support services and hold perpetrators accountable" (campussafety.org/CSA/articles/vawgrants.html, 1999).

The federal government has recently shown much interest in the extent and nature of campus crime. Its interest has manifested into a clear commitment to ensure that victim services, especially sexual victimization services, are available to the campus community and that this availability is communicated to the campus community. Its interest and actions, as we have discussed, will have a lasting impact on the roles and functions of campus law enforcement. An article in the Office for Victims of Crime 1995 publication *National Victim Assistance Academy* (www.ojp.usdoj.gov/ovc/asssist/nvaa/ch21-8cc.html, 1999) nicely summarizes the impact the federal government's interest in campus crime issues has had on campus policing:

> the initial discussion of the Campus Security Act was focused on the reporting of campus crime statistics and confusion about the requirements, the long-term effects of the Act have been far greater. Crime on college campuses and its impact on victims, the college, and the surrounding community has received much needed attention. Many campus law enforcement officials have reported resource increases that have improved security and improved or clarified relationships with local police or sheriff's departments. In addition, many universities have developed or expanded crime victim assistance programs on campus and established more formal ties with off-campus victim assistance programs.

State-Level Legislation.   In 1988, Pennsylvania was the first state to enact legislation requiring colleges and universities to report publicly campus crime statistics; 18 states have since passed legislation designed to require colleges and universities to report campus crime statistics (Fisher, 1995; Griffaton, 1995). Some states have more stringent reporting requirements than those mandated by the Security Act. Wisconsin, for example, requires that rape and acquaintance rape be reported separately (Wis. Stat., 1992). Like the Security Act, all the states except Tennessee require only on-campus victimizations involving students be reported. Tennessee requires its colleges to collect statistics for reported crimes against students that occurred off campus (Tenn. Code Ann., 1994). Some state-level legislation also has less coverage than the federal legislation. Campus crime statistics are gathered in Nevada, for example, only for schools in the University of Nevada system (Nev. Rev. Stat., 1993).

Several states, including California, New York, Washington, and Wisconsin, require that colleges and universities provide services for those victimized on campus. Perhaps the most comprehensive statute is California's law. For example, this law requires "a description of campus resources available to victims as well as off-campus services" (Cal. Educ. Code Ann., 1994, p. 122). The California law also mandates institutions to have policies stating:

> Services available to victims and personnel responsible for providing those services, such as the person assigned to take the victim to the hospital, to refer to the counseling center, and to notify the police, with the victim's concurrence.

Sex-related offenses are the primary focus of New York's campus victim statute. Among other things, it requires that a campus committee be established to annually review security-related policies and procedures including counseling services for victims. The Washington law requires that every new student and staff member be given a security report. This report must include a description of programs regarding counseling as well as a directory of available services. Finally, information with regard to the statutory rights of crime victims must be provided to every new student as required by Wisconsin's law. Additionally, this statement requires that new students also be given information on services for victims of sexual assault or sexual harassment available both on or off campuses.

Research into the Extent and Nature of On-Campus Crime

Researchers have consistently shown that violent and theft victimizations are a young person's game, especially among males. They have also repeatedly shown that rape is primarily committed on young women (Fisher, Sloan, Cullen, & Lu, 1998). Many college students also possess demographic and lifestyle characteristics that researchers have found place people at a higher risk of being victimized: unmarried, low income, and an active social life that includes alcohol and drugs (Fisher et al., 1998).

In light of these research results, the question remains: Do college and university campuses have serious crime problems? Since the answer to this question has implications for campus police and victim services, we need to examine various sources of information: official crime statistics, and results from victimization surveys. As we discuss, the answer to our initial question depends on the type of crime and which sources of campus crime statistics are used. Answers to these questions are important because as we will discuss, they have implications for the types of victim services that are needed on campuses and for the role of the campus police in providing these services or working with other campus officials (e.g., health services, housing) to provide such services.

Official Crime Statistics

Official campus crime statistics are crimes known to the campus police; to be known to the police, the crime has to be reported. If the crime is not reported, then it is not counted in the official crime statistics. Most researchers accept the notion that official crime statistics underestimate the extent of crime as not all crimes are reported to the police.

Past campus crime research and reports have relied heavily on official crime statistics to examine campus crime rates (see Fisher & Sloan, 1995). Frost (1993), for example, using UCR data, reports that between 1985 and 1990, the number of violent crimes reported to university police increased 13%, although the number of property crimes decreased slightly by 0.52%. Bear in mind that the nationwide estimates for violent crime offenses increased 37% and property crimes increased 14% during 1985 and 1990. These results are contrary to those reported by Fernandez and Lizotte (1995). Using data supplied to them from 530 two-year and four-year colleges and universities for 1974 to 1991, they report the rate of violent crime reported to police on campuses decreased 27%.

Estimates of how much crime there is on campuses differ depending on the methodology used to collect the crime statistics. Despite their methodological differences, three reports draw the same general conclusion: property crimes are by far the most frequently occurring crimes on campus. First, in 1997, the Department of Education issued a study done by the National Center for Educational Statistics that examined the extent of campus crime and security at 5,317 institutions in the 50 states, the District of Columbia, and Puerto Rico. Only 44% of schools reported any occurrence on campus of property crimes (burglary and motor vehicle theft) and 26% reported any occurrence on campus of violent crimes (murder, forcible sex offenses, robbery, and aggravated assault). A close look at their results revealed that 80% of these crimes were property crimes (37,780 of the total 47,340).

The second report is from the FBI's Annual Uniform Crime Reporting program *Crime in the United States* (www.fbi.gov/ucr/98cius.html, 1999). In 1998, 489 colleges and universities voluntarily reported their crime statistics to the FBI. Of the 98,754 campus crimes, 96,221 were property crimes (burglary, larceny-theft, motor vehicle theft, and arson). Of the 2,533 violent crimes, 1,588 were aggravated assaults, 535 were robberies, 402 were rapes, and 8 were murders. Compared to the 1997 crime counts, murder, rape, and aggravated assault increased (166.7%, 11.4%, and 2.7%, respectively), and robbery and property crimes decreased (11.4%, and 5.0%, respectively).

Finally, in a 1999 cover story, *The Chronicle of Higher Education* reported official crime statistics for 483 four-year colleges and universities with more than 5,000 students in the United States (Nicklin, 1999). Among institutions reporting data for calendar year 1997, 4,120 violent crimes (murder, forcible and nonforcible sex offenses, robbery, and aggravated assault) were reported to police at these campuses. Those incidents, however, were vastly outnumbered by property crimes such as burglary and car thefts, which totaled 17,904. Arrests for liquor-law violations were significantly higher than arrests for drug-law violations and weapons-law violations (17,624, 7,897, and 951, respectively).

The results from *The Chronicle of Higher Education*'s annual campus crime report support the results from previous campus crime rate research (Bromley, 1992; Sloan, 1994). For example, Sloan (1992a) examined the number of offenses during the 1989-1990 school year for 489 U.S. colleges and universities with at least 3,000 students and which had on-campus housing. He reported that 64% of the offenses involved theft/burglary, 19% involved vandalism, 11% involved drinking and drug-related offenses, 4% involved assaults, and 2% involved violent crimes (e.g., homicide, rape, or robbery).

Results from Victimization Surveys

The second source of information comes from victimization surveys that researchers have administered to a sample of students or faculty. Victimization surveys typically include crime reported and not reported to the campus police. Campus crime researchers have focused on two areas of inquiry: (1) the overall extent and nature of campus victimization, and (2) the sexual victimization of college women. A limited number of general campus crime studies, mostly case studies, using victimization surveys have been done to examine on-campus victimization. Here only a few campus victimization studies have been done at the national level. To some extent, this area of inquiry is in its infancy stages of research.

Although research investigations of the sexual victimization of college women began nearly four decades ago (see Fisher & Cullen, 2000), only recently has a substantial literature examining the rubric of sexual victimization developed. However, little empirical research has been conducted on some forms of victimization, such as stalking.

Results from General Campus Victimization Surveys.   The campus victimization survey suggests that students and faculty are not immune from experiencing on-campus victimization; they are at risk for some types of victimization more so than other types of victimization. For example, based on telephone interviews in the spring of 1994, with 3,472 randomly selected students across 12 four-year colleges and universities, Fisher et al. (1998) reported that 24% of their sample had been victimized on campus at least once since the beginning of the 1993-1994 academic year. Similar to the results from studies using the official crime statistics, Fisher et al. reported that the on-campus theft rate per 1,000 students outnumbered the on-campus violent victimization rate 4 to 1. Students seem particularly at risk for having their property stolen (114.6 victimization rate per 1,000 students), particularly when they are not around to have contact with the offender; personal larceny without contact was the most common type of theft on campus (109.5 per 1,000 students). Among violent crimes, sexual assault[2] and simple assault were the two most common types of victimization (12.7 and 12.1 per 1,000 students, respectively).

Faculty as well are far more at risk of having property stolen than experiencing a personal crime. Using a random sample of faculty members at the University of Cincinnati, Wooldredge, Cullen, and Latessa (1992) reported that 27% of their sample had been victims of on-campus property crimes compared to 5% who had experienced an on-campus personal crime.[3]

---

[2]Sexual assault was defined as "a wide range of victimizations, separate from rape and attempted rape. These crimes include attacks or attempted attacks generally involving unwanted sexual contact between the victim and the offender. Sexual assaults may or may not include force and include such things as grabbing and fondling. Sexual assault also includes verbal threats" (see U.S. Department of Justice, 1998, p. 149).

[3]Personal crimes included robbery, aggravated assault, rape, and assault with a deadly weapon. Property crimes included office burglary, theft of personal property, and damage to office or personal property.

The Sexual Victimization of College Women.   The 1970s and 1980s witnessed many inquiries into the incidence and prevalence of the sexual victimization of college women, including studies on date rape and acquaintance rape, sexual assault, and sexual harassment (see Fisher & Cullen, 2000). For example, evidence indicates that between 8% and 15% of college women have been victims of rape during their college tenure (see Koss, Gidycz, & Wisniewski, 1987; Crowell & Burgess, 1996). Studies of sexual harassment estimate that between 30% to 35% of female students will experience sexual harassment by at least one faculty member during their college tenure (see Belknap & Erez, 1995).

Other research suggests that college women are at a higher risk for rape and other forms of sexual assault than women in the general population (see Belknap & Erez, 1995). For example, Fisher et al. (1998) reported that in their sample, on-campus rape and sexual assault rates for those aged 20 to 24 were 3.3 and 3.1 times higher than the National Crime Victimization Survey rape and sexual assault rates for 1993 and 1994, respectively (that is, rape and sexual assault were 33.9 in their sample compared to 5.7 and 5.0 for persons 20 to 24).[4]

A recent national-level study in the United States of college women who were enrolled in two-year and four-year college and universities with 1,000 or more students sheds some more light on the incidence of sexual victimization among college women. Fisher and Cullen (1998) reported that 11% of their sample experienced some form of sexual victimization during the 1996-1997 academic year.[5] Further, they estimated that 2% of their sample had experienced rape (completed and attempted) on campus during the same academic year. If a school has 5,000 female students enrolled, this means that 100 of the female students would be raped on campus during the school year—an estimate that would alarm any school administrator.

Physical Violence Against College Women.   Researchers have also reported that the incidence of physical violence—hitting, slapping, kicking, or beating—against college women is frequent, especially against those women who are in a dating or an intimate relationship (Crowell & Burgess, 1996). Studies on violence in dating relationships have reported that about three out of five college women know someone who has experienced violence in a dating relationship, and about one in five college women have experienced such violence. Men are also victims of physical violence in dating relationships but not at such high levels (see Belknap & Erez, 1995).

Violence in college dating relationships was also the subject of an inquiry by Sellers and Bromley (1996). In their study, conducted at a large urban university, they found that approximately 22% of their respondents had been the victim of either physical or sexual aggression in their relationship. The majority of violent behaviors involved pushing, grabbing, shoving, and slapping. Although these victimizations may seem minor, they are important from the standpoint of providing interventions to prevent more serious victimizations, as Sellers and Bromley also found that the use of violence increased with the length of the relationship.

---

[4]The U.S. Department of Justice sponsors the National Crime Victimization Survey. It is used to estimate the annual amount of crime victimization in the United States.

[5]Sexual victimization includes completed and attempted rape, completed and attempted sexual coercion, sexual contact with and without force or threat of force, threats of rape and sexual contact with force or threat of force, and sexual harassment.

The Stalking of College Women.   Stalking is a relatively new crime, although by no means a new type of behavior (see Meloy, 1998). The first antistalking law was passed in 1990 in California. Today, all 50 states and the District of Columbia have implemented antistalking laws (Marks, 1997). Congress, too, has passed several pieces of legislation that address stalking (see *the Violence Against Women Act, Title IV of the Violent Crime Control and Law Enforcement Act of 1994*) (Public Law 103-322).

A small body of research has recently appeared that provides data on the incidence of stalking against college women. One school case study estimates that close to 30% of college women were stalked at some point during their lives (see Coleman, 1997; Fremouw, Westrup, & Pennypacker, 1996). Using a six-month reference period with a sample of female students from nine colleges and universities, Mustaine and Tewksbury (1999) reported that 10.5% of the females in their sample said that they had been a victim of behavior that the women defined as "stalking." In the only national-level study of stalking among college women, Fisher et al. (1998) reported that 13% of the sample reported that they met the legal definition of having experienced stalking since school had begun in the fall of 1996 (approximately a seven-month reference period).

The results from these studies illustrate that college and university campuses are not immune from criminal activity. Some types of crimes happen more frequently on campus than other types—property crimes more so than violent crimes. However, some types of physical violence happen often to college women, especially among victims and offenders who know each other. Further, stalking appears to frequently occur among college women.

Although these campus victimization studies provided insights into the extent and nature of on-campus crime, little is known about the services provided to the victims of these crimes. The role of the campus police has clearly evolved over the last 30 years; however, we know very little about the level and types of assistance afforded to campus crime victims. We now turn to highlighting what types of victim services are offered on campus and the role of the campus police in delivering those services.

HOW COLLEGES AND UNIVERSITIES HAVE RESPONDED TO THE NEEDS AND MANDATES FOR VICTIM SERVICES

Various influences have come to bear on colleges and universities to respond to the needs of those victimized on campus as well as students who have been victimized off campus. As we have discussed previously, colleges and universities participating in the federal Title IV financial aid programs are required by federal law to provide victim services, especially to those who have experienced a sexual offense. The Clery Act, for example, requires notification to students of existing on-campus and off-campus counseling, mental health, or other student services available for victims of sexual offenses. Some states also require colleges and universities to provide certain types of victim services. Court decisions, actions by watchdog groups such as SOC, and campus crime estimates that are now available to parents and current and prospective students have influenced schools to respond to the needs of victims.

A review of the existing literature on campus victim services did not yield a wealth of information. However, we were able to find a handful of recent studies that provide a snapshot of the current status of campus victim services. One of the first summaries of campus victim services was provided as part of an overall review of current practices in campus security. Kirkland and Siegel (1994) conducted a nonrandom survey of 32 colleges and universities in an effort to help define exemplary campus security programs. In doing so, they reviewed the offered victim services at these institutions. Their findings suggest that the following services *should be* made available to campus crime victims (Kirkland & Siegal, 1994, p. 8):

- Help in reporting crime to the appropriate authorities.
- Support and advocacy services.
- Psychological and medical services as needed.
- Special help for victims of sexual assault and rape.
- Support for secondary victims, such as friends and roommates of victims.
- Access or referral to support for victim-witnesses in the criminal justice system and in-house judicial procedures.
- Help for academic adjustments, as needed.

How well have schools responded to the need for victim services and federal and state mandates for victim services? Four recent studies shed partial light on this question. First, according to the U.S. Department of Justice (1998, p. 269), many schools "have developed or expanded their crime victim assistance programs on campus and established more formal ties with off-campus victim assistance programs." The 1996 BJS study of campus law enforcement at four-year colleges and universities reported that more than a third of these departments had a special unit or program for victim assistance. This percentage was the smallest percentage among all the other types of special units or programs (e.g., general crime prevention, date rape prevention). They also reported, however, that the majority of large campuses (i.e., 25,000 or more students; n = 57) had a special unit or program for victim assistance.

Second, as the results from the NCES study show (see Table 9–1), a majority of these institutions provided safety/crime prevention presentations to campus groups and published or posted safety reminders (64% and 63%, respectively) (Lewis & Farris, 1997). Fewer institutions provided emergency phone services and victim's assistance services (45% and 33%, respectively). Lewis and Farris also found that the majority of these institutions had instituted or improved all four of these types of services in the last 5 years (82%, 80%, 77%, and 72%, respectively) (Lewis & Farris, 1997, table 18).

As can be seen in Table 9–1, there is considerable variation in the percentage of institutions that offer these services based upon institutional characteristics. A large percent of the four-year public schools offer all four types of programs. For example, 94% of these institutions offer safety/crime prevention presentations, 88% have a program of publishing or posting safety reminders, 79% have emergency phone systems, and 70% offer victim's assistance programs. Even among the four-year schools, variation exists. For example, public four-year colleges were much more likely than private four-year institutions to have victim assistance programs (70% versus 43%).

The National Criminal Justice Association (NCJA), a private organization headquartered in Washington, D.C., conducted the third study on campus victim services. In response to NCJA's solicitation of information, materials describing victim services were received from 38 institutions. Based on a review of the material NCJA staff received, they identified three types of victim services (see Table 9–2) (NCJA, 1999, pp. 29–30). As can

TABLE 9–1   Percent of Post-Secondary Institutions Offering Which Types of Programs By Type of Institution

Institutional Characteristic	Type of Program			
	Safety/Crime Prevention Presentation to Campus Groups	Program of Publishing or Posting Safety Reminders	Emergency Phone System	Victim Assistance Programs
All institutions	64%	63%	45%	33%
Type of Institution				
For profit, less than 2 years	43%	47%	27%	18%
Other, less than 2 years	50%	48%	38%	20%
Public, 2 years	74%	70%	50%	33%
Private, 2 years	52%	54%	38%	29%
Public, 4 years	94%	88%	79%	70%
Private, 4 years	79%	75%	57%	43%

Adapted from U.S. Department of Education (1997), *Campus Crime and Security at Post-Secondary Education Institutions,* p. 37.

be seen, the different types of victim services run along a continuum that begins immediately after the criminal victimization occurs to handling the aftermath of the victimization. For example, emergency services address the immediate needs of victims with respect to their person and/or property and should include coordinated efforts on the part of law enforcement, physical and mental health professionals, victim and social services, and student affairs/services. Advocacy/support services help the victim to deal with the short- and long-term disruptions that a victimization can cause in one's day-to-day life. They also aid the victim in seeking redress from the offender. Counseling services help the victim to move beyond the criminal victimization emotionally and psychologically.

A fourth study, undertaken by Gonzales (1999), examined campus victim services offered by 50 of the largest (based on student enrollment) four-year institutions in the United States (92% response rate, n = 46). She asked the directors of the respective campus police departments if their campuses had a victim assistance office. Supportive of the NCES results, she found that 63% (n = 29) of the institutions had a victim assistance program on campus. The remaining colleges did not have a specific victim assistance office; they either used victim assistance programs in their surrounding communities and/or provided counseling or other assistance with campus resources.

Gonzales (1999) later sent a more detailed survey to the 29 colleges that indicated having victim assistance offices to determine specific services offered to victims. Unfortunately, only slightly over half of those institutions responded. Nonetheless, given the lack of information in this area, it is useful to mention the current practices at some of the

TABLE 9–2   **Type of Victim Service and Types of Services Offered**

Types of Programs Included

Emergency Services

- Crisis intervention (on scene)
- 24-hour hot line or pager numbers available
- Transportation (to hospital, medical appointments, court hearings, etc.)
- Assistance with medical fees
- Shelter/food
- Arrange for repairs after break-in

Advocacy/Support Services

- Intervention with professors
- Intervention with employers
- Intervention with creditors
- Intervention with landlord
- Relocation to new residence hall
- Interpreter for non-English-speaking/English as a second language students, deaf students
- Orientation to student judicial and criminal justice systems
- Written information about student judicial and criminal justice systems and options
- Escort to court hearings and/or judicial hearings
- Escort to medical appointments
- Communication with family if necessary
- Services for faculty and staff
- Victim notification regarding offender release, case status, etc.
- Assistance preparing victim impact statements, documentation for restitution
- Assistance with insurance, victim compensation claims
- Ensure prompt return of property

Counseling

- Refer victims to off-campus professional counseling services
- Arrange for/refer to professional counseling with staff or campus health services
- Outreach/counseling to secondary victims (victim's friends, classmates, resident assistants/directors, professors, campus community)
- Mediation between victim and offender

Adapted from the 1999 National Criminal Justice Association report on services for campus crime victims.

TABLE 9–3   Percentage of Institutions Providing Various Victim's Services

Type of Service	Percent of Institutions
Letters to instructors regarding victims	93%
Accompany to student affairs office	87%
Assist filing police reports	73%
Accompany to prosecutor and medical appointments	67%
Arrange or provide for debriefing	60%
Assist filing instructions	47%
Provide safe housing	27%

Adapted from Gonzales (1999), *A Review of Victim Services on College Campuses.*

largest institutions in our country. Table 9–3 shows the services typically provided by the victim assistance offices at those campuses.

Taken together, these studies document that many colleges and universities provide some form of victim services and offer insight into the different ways in which campus police have responded to the safety, security, and victim needs of the campus community. Two case studies will help to show what innovative victim assistance services are currently in practice.

Case Studies

Michigan State University.   One aspect of victim services at Michigan State University deals with sexual assault. The Department of Public Safety at Michigan State University has long recognized the importance of providing assistance to crime victims in general and sexual assault victims in particular. Some years ago the Michigan State Department of Public Safety worked closely with various faculty and representatives of the campus sexual assault program to develop a "guarantee" to their campus members regarding their commitment to the problem of sexual assault. According to Benson (1992, p. 26) the department felt that "a written guarantee to the sexual assault victims might accomplish some of the following objectives:"

- Stimulate awareness, discussion, and reporting of the "hidden" and often unreported crime of acquaintance rape.
- Help put sexual assault victims more at ease in accessing and working with their local police officers.
- Emphasize a strong organizational commitment by the Department of Public Safety (DPS) to sexual assault victims.
- Enable DPS to reach out to sexual assault victims, who have already suffered much trauma, to provide help so that they are not further victimized by the criminal justice system itself.

The actual "guarantee" is shown in Figure 9–1.

Sexual assaults, including date/acquaintance rape, are a very serious concern of DPS. If you feel you are the victim of a sexual assault on campus, your Department of Public Safety will guarantee you the following.

1. We will meet with you privately, at a place of your choice in this area, to take a complaint report.
2. We will not release your name to the public or to the press.
3. Our officers will not prejudge you, and you will not be blamed for what occurred.
4. We will treat you and your particular case with courtesy, sensitivity, dignity, understanding, and professionalism.
5. If you feel more comfortable talking with a female or male officer, we will do our best to accommodate your request.
6. We will assist you in arranging for any hospital treatment or other medical needs.
7. We will assist you in privately contacting counseling, safety, advising, and other available resources.
8. We will fully investigate your case, and will help you to achieve the best outcome. This may involve the arrest and full prosecution of the suspect responsible. You will be kept up-to-date on the progress of the investigation and/or prosecution.
9. We will continue to be available for you, to answer your questions, to explain the systems and processes involved (prosecutor, courts, etc.), and to be a listening ear if you wish.
10. We will consider your case seriously regardless of your gender or the gender of the suspect.

If you feel you are a sexual assault victim, call your Department of Public Safety at 355-2221, and say you want to privately make a sexual assault complaint. You may call any time of day or night.

   If we fail to achieve any part of the above guarantee, the Director of Public Safety, Dr. Bruce Benson (355-2223), will meet with you personally to address any problems. DPS wants to help you make the MSU campus safe for students, faculty, staff, and visitors.

FIGURE 9–1   Department of Public Safety sexual assault guarantee. (Adapted from *Campus Law Enforcement Journal,* November/December, 1992.)

University of South Florida.   The University of South Florida in Tampa is the second largest in the state of Florida and serves a daily population of over 40,000 persons. The Victim Advocacy Program on this campus was established after a well-publicized series of sexual assault allegations against a star basketball player. Unfortunately, this case was improperly handled by the university, which led to a review by a committee appointed by the Florida Board of Regents. Following that review, the Victim Advocacy Program was established in order to provide a full range of services to victims of campus crime. The Victim Advocacy Program reports to an administrative vice president, separate from the university police and from the Office of Student Affairs. It is believed that the level of autonomy granted to the Victim Advocacy Program is useful in encouraging victims of violent crime to report the crime, even if they do not wish the police to be involved. The Victim Advocacy Program then *encourages* victims to make reports to the police but only if they are so willing. On this campus, the Victim Advocacy Program and the police department also work closely on mutual training, crime prevention programs, and coordination with the media. This cooperative relationship helps to assure that campus crime victims receive the best possible services (Poole, 1995).

CHALLENGES TO CAMPUS POLICE IN PROVIDING QUALITY VICTIM SERVICES

There are many challenges facing campus police and other campus officials who must and do routinely provide services to campus crime victims. Each campus is unique with respect to its crime problems. As such, the needs of the victims and the services that can address those needs require a tailored approach to each respective campus. Like generic crime prevention, which the research has consistently shown not to be effective in reducing crime, generic victim services may be ineffective as well (Fisher et al., 1998). To date, however, there is little, if any, research that has evaluated the quality or effectiveness of the campus police in providing victim services or their relationship with other service providers.

This area of campus police research is ripe for inquiry because there are several important issues that we discussed at length in this chapter that have implications for campus policing and victim services. First, many campus police have evolved from their watchman–security guard roles to law enforcer–service provider roles. These "new" roles and functions of the campus police in providing victim services, especially in the progressive campus departments that have adopted community policing principles, have implications for the future development of new roles that campus police departments can have in providing victim services. Second, campus safety, security, and victim services are policy issues that Congress, state legislatures, the media, the courts, and watchdog organizations such as SOC have maintained an active interest in addressing. Third, research has shown that all types of personal and property crime on campus are day-to-day realities that the campus community must frequently face. There are some types of crimes (theft, for example) that members of the campus community must face more often than other types of crime. This is not to suggest that more serious types of violence are rare. They do happen on campus—and for sexual victimization, at rates higher than rates for comparable age groups (Fisher et al. 1998). Fourth, the research to date suggests a consistent picture of campus police; the majority of campus police departments are providing some form of victim services.

As campus police become more involved in developing and providing victim services and working with other victim service providers, there are several challenges that they face. First, senior campus administrators must acknowledge that serious violent crimes can and do occur within the college setting—on campus and off campus—both near and far. This is an important first step in developing and implementing comprehensive strategies to prevent victimizations and to assist victims after the incident has occurred (Belknap & Erez,

1995; Benson, Carlton, & Coodhart, 1992; Roark, 1987). Too often this acknowledgment is made only after a tragic crime of violence on a college campus or among college students off campus. Some crimes require follow-up by a variety of service providers both on and off campus. For example, it is generally accepted that victims of crimes such as stranger rape or domestic violence may need short-term and long-term crisis intervention services as well as other psychological, legal, and financial assistance. If the crime is campus-related, the assistance may be provided by campus or community resources or sometimes by both working together (Tomz & McGillis, 1997). Establishing and maintaining communication links between campus police and victim assistance agencies outside the police department both on campus and off campus is essential to coordinating the management of victim cases.

Second, the campus police should work to reduce barriers to accessing available victim services. Here among the most daunting challenges is getting students to report their victimization. A national study of 3,472 students at 12 randomly selected four-year institutions reported that 76% of on-campus personal victimizations (e.g., rape, sexual assault, robbery, and assault) and 75% of thefts were unreported to the campus police (Sloan, Fisher, & Cullen, 1997). Some institutions have implemented practices that make reporting easier by offering "silent reporting" in which reporting of an incident can be done using the Internet and/or "proxy reporting" where a third party can report an incident. However, without any report of an incident, campus police may be at a loss or totally ineffective in using their victim resources to provide needed victim services. Getting students (and faculty and staff) to report incidents may help campus police to determine what types of crimes happen where, when, and under what circumstances. This understanding of the incident and its causes will then aid in the development of victim services that are tailored to the needs of the victims and contribute to the effective delivery and impact of these services.

Third, as we have highlighted, a relatively high incidence of stalking among college women has been documented by researchers (Fisher et al., 1998). Campus police and other victim service providers may be challenged by this "new" crime. Little is known about the effects of stalking among college students. Studies among the members of the general population show that stalking victims may experience negative physical and psychological effects both short term and long term (see Meloy, 1998). Stalking victims may require different types of services than what institutions currently have in place for sexual assault or sexual harassment victims.

Fourth, "underserved victims"—including male victims of sexual assaults, property crime victims, victims with disabilities, commuter students, victims of hate/bias crime, international students, and students of diverse culture—may pose needs that do not fit neatly into currently offered victim services (NCJA, 1999). Campus police may not be adequately trained and/or have the resources to address the needs of the underserved victims. The NCJA report (1999) highlighted that resources should be used to assess the needs of these victims, improve outreach to them, and eliminate barriers for them to access services.

We are aware of the scarcity of funding when suggesting more services. We are not advocating more and more resources to victim services in an effort to "do something for every victim." Instead, we are suggesting that campus police and administrators take a more sustained and judicious approach in which they pay more attention to how victim resources should be allocated given the types of crimes that happen on their respective campuses and how best to make victim services more effective to each victim of any type of crime.

Campus police and administrators cannot ignore these challenges to providing quality victim services as they compete to attract and recruit students, faculty, and staff; maintain enrollments; reduce liability; and fall under the scrutiny of concerned parents and students, campus-safety interest groups, and state and federal legislation. The challenge for campus police and administrators, then, is to develop victim services that are shaped by the specific victimization patterns besetting their respective institutions. In this way, campus police can be more effective providers and coproviders of victim services and be more strategic in the types of victim services that they provide and in doing so, hopefully provide quality services to victims of any type of crime.

REFERENCES

BELKNAP, J., & EREZ, E. (1995). The victimization of women on college campuses: Courtship violence, date rape and sexual harassment. In B.S. Fisher & J.J. Sloan (Eds.), *Campus crime: Legal, social and policy perspectives* (pp. 156–178). Springfield, IL: Charles C. Thomas.

BENSON, B. (1992). The public safety "sexual assault guarantee" at Michigan State University. *Campus Law Enforcement Journal,* November-December, 22, 26–28.

BENSON, B.L. (1993). Community policing works at Michigan State University. *Journal of Security Administration, 16*(1), 43–52.

BENSON, D., CHARLTON, C., & COODHART, F. (1992). Acquaintance rape on campus. *Journal of American College Health, 40,* 157–165.

BRANTINGHAM, P., BRANTINGHAM, P., & SEAGRAVE, J. (1995). Crime and fear at a Canadian university. In B.S. Fisher & J.J. Sloan (Eds.), *Campus crime: Legal, social and policy perspectives* (pp. 123–155). Springfield, IL: Charles C. Thomas.

BROMLEY, M.L. (1992). Campus and Community Crime Rate Comparisons: A Statewide Study. *Journal of Security Administration, 15*(2): 49–64.

BROMLEY, M.L. (1998). Campus crime victims: Legal, statutory and administrative responses. In L.J. Moriarty & R.A. Jerin (Eds.), *Current issues in victimology research* (pp. 93–110). Durham, NC: Carolina Academic Press.

BROMLEY, M.L. (1996). Policing our campuses: A national review of statutes. *American Journal of Police, 15*(3), 1–22.

BUREAU OF JUSTICE ASSISTANCE. (1997). *Crime prevention and community policing: A vital partnership.* Washington, DC: United States Department of Justice.

CARRINGTON, F. (1991). Campus crime and violence: A new trend in crime victims' litigation. *The Virginia Bar Association Journal, XVII*(1), 50.

CASTELLI, J. (1990, November 4). Campus crime 101. *The New York Times Education Life,* p. 1.

COLEMAN, F.L. (1997). Stalking behavior and the cycle of domestic violence. *Journal of Interpersonal Violence, 12,* 420–432.

CROWELL, N.A., & BURGESS, A.W. (1996). *Understanding violence against women.* Washington, DC: National Academy Press.

ESPOSITO, D., & STORMER, D. (1989). The multiple roles of campus law enforcement. *Campus Law Enforcement Journal, 19*(3), 26–30.

FERNANDEZ, A., & LIZOTTE, A. (1995). An analysis of the relationship between campus crime and community crime: Reciprocal effects? In B.S. Fisher & J.J. Sloan (Eds.), *Campus crime: Legal, social and policy perspectives* (pp. 79–102). Springfield, IL: Charles C. Thomas.

FISHER, B., SLOAN, J.J., CULLEN, F., & LU, C. (1998). Crime in the ivory tower: The level and sources of student victimization. *Criminology, 36*(3), 671–710.

FISHER, B.S. (1995). Crime and fear on campus. *The Annals of the American Academy of Political and Social Sciences, 539,* 85–101.

FISHER, B.S., & CULLEN, F.T. (2000). Measuring the sexual victimization of women: Evolution, current controversies, and future research. in NIJ2000 series, Volume 4. Washington, DC: NIH.

FISHER, B.S., & CULLEN, F.T. (1998). The sexual victimization of college women: Findings from a national-level study. Final report submitted to the National Institute of Justice.

FISHER, B.S., & SLOAN, J.J. (1995). *Campus crime: Legal, social and policy perspectives.* Springfield, IL: Charles C. Thomas.

FREMOUW, W.J., WESTRUP, D., & PENNYPACKER, J. (1996). Stalking on campus: The prevalence and strategies for coping with stalking. *Journal of Forensic Sciences, 42,* 666–669.

FROST, G.A. (1993). Law enforcement responds to crime on campus. *Journal of Security Administration, 16*(1), 21–30.

GELBER, S. (1972). *The role of campus security in the college setting.* Washington, DC: United States Department of Justice.

GONZALES, M. (1999). *A review of victim services on college campuses.* Unpublished research project, University of South Florida, Tampa.

GRIFFATON, M.C. (1995). State-level initiatives and campus crime. In B.S. Fisher & J.J. Sloan (Eds.), *Campus crime: Legal, social and policy perspectives* (pp. 52–73). Springfield, IL: Charles C. Thomas.

JOHNSON, R. (1995). Implementing community policing: One university's experience. In *Community policing on campus.* Hartford, CT: International Association of Campus Law Enforcement Administrators.

KALETTE, D. (1990, September 14). Colleges confront liability. *U.S.A. Today,* p. 6A.

KIRKLAND, C., & SIEGEL, D. (1994). *Campus security: A first look at promising practices.* Washington, DC: U.S. Department of Education.

KOSS, M.P., GIDYCZ, C.A., & WISNIEWSKI, N. (1987). The scope of rape: Incidence and prevalence of sexual aggression and victimization in a national sample of higher education students. *Journal of Consulting and Clinical Psychology, 55*(2), 162–170.

LANIER, M. (1995). Community policing on university campuses: Tradition, practice, and outlook. In B.S. Fisher & J.J. Sloan (Eds.), *Campus crime: Legal, social and policy perspectives* (pp. 246–264). Springfield, IL: Charles C. Thomas.

LEDERMAN, D. (1993, January 20). Colleges report 7,500 violent crimes on their campuses in first annual statements required under federal law. *The Chronicle of Higher Education,* pp. A32–A43.

LEDERMAN, D. (1994, February 3). Crime on the campus. *The Chronicle of Higher Education,* p. A33.

LEDERMAN, D. (1994, March 9). Weapons on campus. *The Chronicle of Higher Education,* p. A33.

LEWIS, L., & FARRIS, E. (1997). *Campus crime and security at post-secondary institutions* (NCES 97-402). Washington, DC: U.S. Department of Education, National Center for Education Statistics.

LIVELY, K. (1997, March 21). Campus drug arrests increased 18 percent in 1995: Reports of other crimes fell. *The Chronicle of Higher Education,* p. A44.

LIVELY, K. (1998, May 8). Alcohol arrests on campuses jumped 10% in 1996; Drug arrests increased by 5%. *The Chronicle of Higher Education,* p. A48.

LIVELY, K. (1996, April 26). Drug arrests rise again. *The Chronicle of Higher Education,* p. A37.

MARKS, C.A. (1997).The Kansas stalking law: A "credible threat" to victims. *Washburn Law Journal, 36,* 468–498.

MATHEWS, A. (1993, March 7). The campus crime war. *The New York Times Magazine,* 38–47.

MELOY, J.R. (1998). *The psychology of stalking: Clinical and forensic perspectives.* San Diego, CA: Academic Press.

MILLER, M., HEMENWAY, D., & WECHSLER, H. (1999, July). Guns at college. *The Journal of American College Health, 48,* 7–12.

MUSTAINE, E.E., & TEWKSBURY, R. (1999). A routine activity theory explanation for women's stalking victimizations. *Violence Against Women, 5*(1), 43-62.

NATIONAL CRIMINAL JUSTICE ASSOCIATION. (1999). *Call to the field: A summary of the national assessment report.* Washington, DC: Author.

NICHOLS, D. (1986). *The administration of public safety in higher education.* Springfield, IL: Charles C. Thomas.

NICHOLS, D. (1997). *Creating a safe campus: A guide for college and university administrators.* Springfield, IL: Charles C. Thomas.

NICKLIN, J.L. (1999, May 28). Colleges report increases in arrests for drug and alcohol violations. *The Chronicle of Higher Education,* p. A39.

ORDOVENSKY, P. (1990, December 3). Students easy prey on campus. *U.S.A. Today,* p. 1A.

PEAK, K.J. (1995). The professionalization of campus law enforcement: Comparing campus and municipal law enforcement agencies. In B.S. Fisher & J.J. Sloan (Eds.), *Campus crime: Legal, social and policy perspectives* (pp. 228–245). Springfield, IL: Charles C. Thomas.

POOLE, M. (1995). The evolution of victim advocacy at the University of South Florida. *Campus Law Enforcement Journal,* January-February , 25, 27–28.

POWELL, J. (1981). The history and proper role of campus security. *Security World, 8*(1), 18–25.

POWELL, J., PANDER, M., & NIELSEN, R. (1994). *Campus security and law enforcement,* (2nd ed.). Boston: Butterworth-Heinemann.

REAVES, B., & GOLDBERG, A. (1996). *Campus law enforcement agencies, 1995.* Washington, DC: United States Department of Justice.

ROARK, M. (1987). Preventing violence on campus. *Journal of Counseling and Development,* March (65), 367–371.

SELLERS, C.S. , & BROMLEY, M.L. (1996). Violent behavior in college student dating relationships: Implications for campus service providers. *Journal of Contemporary Criminal Justice, 12*(1): 1–27.

SLOAN, J.J. (1992a). Campus crime and campus communities: An analysis of campus police and security. *Journal of Security Administration, 15*(2), 31–45.

SLOAN, J.J. (1992b). The modern campus police: An analysis of their evolution, structure, and function. *American Journal of Police, 11*(1), 85–104.

SLOAN, J.J. (1994). The correlates of campus crime: An analysis of reported crimes on university campuses. *Journal of Criminal Justice, 22*(1), 51 62.

SLOAN, J.J., FISHER, B.S., & CULLEN, F.T. (1997). Assessing the Student Right to Know and Campus Security Act of 1990. *Crime and Delinquency, 43*(2), 148–168.

SMITH, M.C. (1989). *Campus crime and campus police: A handbook for police officers and administrators.* Asheville, NC: College Administration Publications, Inc.

SMITH, M.C. (1995). Vexations victims of campus crime. In B.S. Fisher & J.J. Sloan (Eds.), *Campus crime: Legal, social and policy perspectives* (pp. 25–37). Springfield, IL: Charles C. Thomas.

SMITH, M.C., & FOSSEY, R. (1995). *Crime on campus: Legal issues and campus administration.* Phoenix, AZ: American Council on Education and The Oryx Press.

TOMZ, J., & MCGILLIS, D. (1997). *Serving crime victims and witnesses.* Washington, DC: U.S. Department of Justice.

U.S. DEPARTMENT OF HEALTH, EDUCATION, AND WELFARE, EDUCATION DIVISION, NATIONAL CENTER FOR EDUCATION STATISTICS. (1997). *Digest of education statistics.* Washington, DC: Author.

U.S. DEPARTMENT OF JUSTICE, OFFICE OF JUSTICE PROGRAMS, OFFICE FOR VICTIMS OF CRIME. (1998). *New directions from the field: Victims' rights and services for the 21st century.* Washington, DC: Author.

WEBB, J. (1975). The well-trained professional university police officer: Fact or fiction? *FBI Law Enforcement Bulletin, 44*(1), 26–31.

WHITAKER, L.C., & POLLARD, J. (1993). *Campus violence: Kinds, causes, and cures.* Binghamton, NY: The Haworth Press, Inc.

WOOLDREDGE, J.D., CULLEN, F.T., & LATESSA, E.J. (1992). Victimization in the workplace: A test of routine activities theory. *Justice Quarterly, 9*(2), 325–335.

INTERNET LINKS

www.campussafety.org/CSA/articles/vawgrants.html (1999)

www.campussafety.org/cas/right.html (1999)

www.fbi.gov/ucr/98cius.html (1999)

www.ojp.usdoj.gov/ovc/assist/nvaa/ch21-8cc.html (1999)

www.soconline.org (1999)

www.uab.edu.police/housing.html (1999)

CASES AND STATUTES

Cases

- *Duarte v. State of California et al.,* 151 CAL. RPT. 727 (Cal. App. 1979)
- *Gross v. Family Services Agency, Inc.,* 716 So. 2d 337 (Fla. Dist. Ct. App. 1998)
- *Johnson v. Washington,* 894 P.2d 1366 (Wash. Ct. App. 1995)
- *Miller v. State of New York,* 62 N.Y.2d 506, 478 N.Y.S.2d 829, 467 N.E.2d 493 (1984); as to damages see 110 A.D.2d 627, 487 N.Y.S.2d 115 (1985)
- *Mullins v. Pine Manor College et al.,* 449 NE 2d 331 (Mass. 1983)
- *Nero v. Kansas State University,* 861 P.2d 768 (Kan. 1993)
- *Peterson v. San Francisco Community College District,* 685 P.2d 1193 (al. 1984)

Statutes

- Cal. Educ. Code Ann., §§67380, 67390 to 67393, 94380 (1994)
- N.Y. Educ. §6450 (McKinney 1985 and Supp. 1994)
- Wash. Rev. Code §28B.10.569 (Supp. 1993)
- Wis. Stat. §36.11(22) (West Supp. 1992)

10

Policing and Victims

Children and Others

M.L. Dantzker

Laura J. Moriarty

INTRODUCTION

Victims. . .they come in all shapes and sizes, colors, and genders. As the previous nine chapters have demonstrated, there is much to learn and to know about dealing with victims. A main point is that there is always a victim and that the victim deserves as much, perhaps even more, attention than the offender or any other actor in the criminal justice system. Nowhere is this more true than when the victim is a child.

Just prior to starting this chapter, a news story out of Mount Morris Township, Michigan, was being aired: a 6-year-old boy had been accused of fatally shooting a 6-year-old female classmate. Our natural inclination is to empathize with the victim and her family; but what about the boy? According to the Genesee County Prosecutor, the "boy comes from a very troubled home . . . he is really a victim of a drug culture and a house that's really in chaos" (Pickler, 2000, p. 1). The prosecutor is also quoted as saying about the boy, "He is a victim in many ways. . . It is very sad, we need to put our arms around him and love him" (Pickler, 2000, p. 2). The question is, who will meet this need? Where in the system is the individual or organization who will look after the well-being of this young boy or any child victim? This chapter introduces one such entity whose sole purpose is to protect and be an advocate for the child victim. Furthermore, we will look at special prosecutorial units, and summarize the state of policing and victims as denoted throughout this text.

CHILD VICTIMS

What is a child victim? A child victim is an individual aged 17 and under, who has been victimized in some form or fashion, be it psychological, sociological, or physical. Although there are probably many children who have been a victim of all three areas, perhaps the ones we are most cognizant of are those who are victims of a physical nature. At least, it is victims in this group who we can more readily count.

According to the executive summary of a report on child victimizers, between 1976 and 1994, 37,000 children were murdered, the largest percentage of which were murdered by family members (Greenfeld, 1996). Other characteristics of child victims reported were:

- Three in 4 child victims of violence were female,
- Children aged 12 to 17, while making up 10% of the population, were 22% of those victimized by violent crime,
- One third of the victimizers committed their crime against their own child,
- Four in 10 child victims of violence suffered either a forcible rape or another injury, and
- More than half the violent crimes committed against children involved victims age 12 or younger (Greenfeld, 1996, pp. 1–2).

Who is the victimizer of the young? The majority of violent victimizers against children are male, almost 25% are 40 years old and older, nearly 70% serving time are white, and almost one third of the child-victimizers have no previous arrest record (Greenfeld, 1996). Perhaps one of the most disconcerting statistics is that six of seven victimizers reported knowing the victim in some manner; they were not strangers. If a child cannot trust those who are close to him or her, then who will protect the child when he or she is victimized?

PROTECTING THE CHILD

When a child is victimized, the first line of protection should be the parents, but what happens when the victimizer is the parents? Then society must step in, which is initially done in the form of a police officer. Unfortunately, policing is extremely limited as to how much it can do for child victims and most often must turn to other social agencies to step in. One of the most recognized social protectors of children is the Department of Children and Family Services (DCFS) or a similar type organization. This social protector is looked upon to provide shelter and guardianship over the child victim whether it is in the form of a youth home or center or a foster home. This entity may be able to take care of the child's immediate physical needs and comforts, but what about when it comes to the courts? What happens when the courts come into the picture? Who will represent the child? As Moriarty and Kenworthy (1998, p. 142) noted, "Children need separate court representation when a petition has been filed alleging neglect, abandonment, mistreatment, sexual, and/or physical abuse. This representation allows the court to hear what is in the best interest for the child." One of the most prominent and growing entities in this country for representing the child in court is the Court Appointed Special Advocate (CASA) program.

WHAT IS CASA?

Because each year thousands of children find themselves in court through no fault of their own, becoming further victimized at the hands of an overburdened child welfare system, there is a tremendous need for them to have an advocate. Stepping into this role is CASA, a program that was "created in 1977 to make sure that the abuse and neglect that these children originally suffered at home doesn't continue as abuse and neglect at the hands of the system" (CASA of Hidalgo County, Texas, 1999, p. 1). The CASA program is primarily comprised of trained community volunteers who are "appointed by a juvenile or family court judge to speak for the best interest of the children who are brought before the court" (CASA, 1999, p. 1). The role of these advocates is threefold:

1. to serve as a fact-finder for the judge by thoroughly researching the background of each assigned case;
2. to speak for the child in the courtroom, representing the child's best interests; and
3. to continue to act as a "watchdog" for the child during the life of the case, ensuring that it is brought to a swift and appropriate conclusion (CASA, 1999).

The CASA program, since its inception in 1977,

has had a dramatic impact on the nation's court system. There are now 843 CASA programs across the country, including Washington, D.C. and the U.S. Virgin Islands. New programs start up at a rate of two per month. Research shows these programs utilize more than 42,400 volunteers, who help an estimated 25 percent of the nation's abused and neglected children in dependent proceedings. In 1997, they worked with approximately 172,000 children. (CASA, 1999, p. 2)

Because CASA programs may be known by different names, police agencies should learn if there is such a program in their jurisdiction and what it is called. For example, in San Diego, California, the program is called Voices for Children (CASA, 1999, p. 2). In Virginia it is called the Virginia Court Appointed Special Advocate Program. In North Carolina it is called the North Carolina Guardian Ad Litem program (Moriarty & Kenworthy, 1998). "One of the largest programs in the nation is the Florida Guardian Ad Litem program, managed by state government" (CASA, 1999, p. 2). A national association was established in 1982 and is based in Seattle, Washington.

Are these programs effective? "Preliminary findings show that children who have been assigned CASA volunteers tend to spend less time in court and less time within the foster care system than those who do not have CASA representation. Judges have observed that CASA children also have better chances of finding permanent homes than non-CASA children" (CASA, 1999, p. 5). Despite the preliminary status of current information about the effectiveness of CASA programs, it appears to be one of the most prolific attempts to date to protect and represent the child victim. Although there may be other types of child advocacy organizations, "CASA is the only program where volunteers are appointed by the court to represent a child's best interests" (CASA, 1999, p. 5).

Obviously children are just one set of victims that the police and criminal justice system must contend with. Yet, they are one of the most precarious groups to deal with because of age and shortcomings in representation. Nevertheless, programs such as CASA are being developed to help in this area. The police would do well to use such groups to their advantage—an advantage that also exists in specially developed units.

SPECIAL UNITS

As discussed briefly in Chapter 8, police departments often establish LEVA (law enforcement victim assistant) units. The purpose of these units is to provide services to victims where police officers work directly with either victims of all crimes or victims of specific crimes. Additional victim services have been developed in courts either through the district attorney's office, the commonwealth attorney's office, or the prosecutor's office (depending on what the state calls the office). Many courts have established specialized units to handle specific crime victims. For example, it is common to find a domestic violence unit or a sexual assault unit where the victim advocates have specialized training, expertise, and experience with victims of the specific crime.

Finn and Lee (1987, p. 16) provide a list of services that experts agree should be part of any victim/witness program. These services include six major areas: emergency services, counseling, advocacy and support services, claims assistance, court-related services, and system-wide services. Within each category are multiple services that all victim units or victim programs should strive to deliver. Finn and Lee include medical care, shelter, security repair, and direct financial assistance in the category of emergency services. Twenty-four-hour hot lines, crisis intervention, follow-up counseling, and mediation are included in counseling. Personal advocacy, employer intervention, landlord intervention, property return, intimidation protection, paralegal/legal counsel, and referral are included in advocacy and support services. Insurance claims aid, restitution assistance, and compensation assistance are included in claims assistance. Court-related services include the longest list of services including witness reception area, court orientation for adults and children, notification, witness alert, transportation, child care, escort to court, and victim impact statements. Within system-wide services are public education, legislative advocacy, and training.

In Richmond, Virginia, the commonwealth attorney's office has six specialized victim units/victim advocates. One unit works only with juveniles who have been victims of any type of crime. There is a special unit working with victims of robbery, aggravated assault, simple assault, and domestic violence. There is also a unit that works with homicide victims' relatives. The sheer number of cases for each crime was the impetus for developing special units. Additionally, the services provided to victims of different crimes vary with the crime. For example, homicide victims' families need assistance with funeral preparations including expenses. Aggravated assault victims often need financial assistance with medical bills. Robbery victims need counseling.

All the victims, regardless of type of crime, must be informed about the criminal justice system. At a minimum the units provide this information, in addition to escorts and accompaniment to court procedures. The units strive to meet the other service needs identified by Finn and Lee (1987).

Like the programs for children, these special units provide policing with additional assistance to deal with one of the most important, yet often overlooked elements of our system, the victim. This text demonstrates just how important these victims are, and how the police can assist them.

CONCLUSION

The purpose of this reader is to identify potential areas of conflict between police officers and victims. The previous nine chapters address topics that quite possibly could cause conflict between the two, if both parties are not educated about the expectations and experiences of the other. Ideally, this book should help avoid some of these conflicts by providing police officers with a better understanding of what victims expect from the police and how to react to victims of crime.

A basic fundamental understanding of the subject matter is essential before a public agency, such as the police, can begin to provide assistance. With this in mind, Chapter 1 begins with an exploration and conceptualization of the term *victimology*. Matt Robinson challenges the reader to expand traditional definitions of victimology, providing a discussion of how such an expanded definition affects policing. Instead of relying on the traditional definition of victimology, Robinson does what no other author has yet done. He defines victimology in a broader, more complete definition.

Robinson suggests that victimization includes "any act which produces financial or physical harm and which is committed intentionally, negligently, recklessly, or knowingly." Furthermore, he advises that by adopting this broader definition, policing would need to reorganize its priorities and focus its resources on a wider group of victims. Ultimately, it is offered that there would be a need for policing to redistribute how time is used and allocation of resources.

The next two chapters are less theoretical than Chapter 1 and focus on explaining a specific problem, providing practical applications or advice to address them. For example, Chapter 2 examines how the police should treat victims of crime. It clearly articulates what victims expect from the police and how police can meet the expectations of the crime victim. For example, the author, Amie Scheidegger, advises that police officers should be aware of the stages of reaction a victim will go through (impact, recoil, and reorganization) and how, depending on the stage the victim is in at the time of police contact, it will influence the victim's response to the police. Furthermore, Scheidegger provides practical advice on how the police should work with distinct victim populations such as the elderly, disabled (hearing and seeing impaired), the ethnically diverse, and secondary victims.

In Chapter 3, Peter Mercier addresses a common problem in policing —how to get victims to report crimes. Mercier examines the criminological literature summarizing what we know works in increasing victim reporting. He also points out why victims often do not report crime. He does this in order to learn more about how to get victims to report more regularly. Finally, Mercier provides additional creative and innovative methods to further increase reporting. For example, he suggests that training programs that increase sensitivity and awareness of victims' issues, prompt property-return procedures, and periodic updates to victims about their cases may increase reporting of victimizations.

The next section of the reader examines specific crimes, addressing how the police should handle such incidents. The subsequent three chapters focus on one category of crime—sexual assault, child welfare, and domestic violence. The authors of these chapters, Tracy Woodard Meyers, Janet Hutchinson, and Denise Kindschi Gosselin, have many years of experience working with the police in relation to these specific crimes. The chapters represent the synergy between their practical experience and scholarly knowledge, providing the reader with invaluable techniques for working with such victims.

Meyers, a family therapist, uses her clinical experience to explain how the police should interview victims of sexual assault. She provides advice regarding successful interviewing strategies. Of particular importance is that police officers must overcome the common myths associated with sexual assault, especially the stereotyping and personal biases. Meyers also notes that police officers need to attain a better understanding of the stress reactions to this traumatic event. Furthermore, because the interview is extremely important, she advises that a successful interview includes knowing when and where to interview the victim, who should conduct the interview, the importance of personal assessment, and how to build rapport. Ultimately, Meyers advises that those who investigate sexual assaults be specifically trained and educated in how to deal with these victims so that the individual does not end up being "revictimized."

Hutchinson, with over 20 years of consulting experience with state and local child welfare cases in 48 of the 50 states, explains how the police should investigate child welfare cases including dependent, neglected, and abandoned children, and children who have been physically and sexually assaulted. She articulates the issues involved in developing and maintaining successful interdisciplinary investigations of child abuse and neglect. Team members representing different disciplines bring very different perspectives of their responsibilities to the group effort—perspectives that often lead to misunderstandings that undermine multidisciplinary cooperation on child abuse and neglect cases. Hutchinson examines these goal conflicts, and the advantages and disadvantages of police involvement in child abuse and neglect cases. She also explores strategies that have been successfully used for developing and maintaining viable multidisciplinary teams that include police in the investigations.

Gosselin, currently a Massachusetts state police officer, details how to interview victims of domestic violence. She discusses the four goals of interviewing: victim identification, risk assessment, evidence gathering, and outcome determination. Especially important in this discussion as it relates to victim identification is overcoming several false assumptions (e.g., the injured person is the victim). This is followed by explaining various methods of interviewing including traditional interviewing, behavioral approaches to interviewing, the kinesic interviewing techniques, the Reid technique of interviewing and interrogation, cognitive interviewing, and forensic interviewing.

The last section of the reader focuses on resources. Police officers in general have a limited awareness of resources available to assist victims. These chapters are included to provide practical information regarding victim resources in an effort for police to be more responsive to victims.

At a minimum the police must understand the rights afforded victims of crime. Jerin provides a detailed modern history of the victims' rights movement beginning in 1964 through 1998. He highlights historically the major achievements in the victims' rights movement to enact legislation that provides victims with certain rights such as compensation, protection, and being kept informed. Figure 7-1 provides an example of one state's summary of victim and witness rights afforded to its citizens.

Moriarty and Diehl expand on Jerin's work. They begin by looking at the things that can be done by police officers to address victims' needs. These include safety and security, ventilation and validation, and prediction and preparation. Furthermore, they reinforce previous chapters' positions that crime victims really want a caring and compassionate response to their victimizations. This is followed by the provision of a detailed list of national resources available to assist victims such as Parents of Murdered Children and Mothers Against Drunk Driving. They also include statewide and local services, such as the office of the state attorney general. Although the focus is on one state—North Carolina—police officers can easily adapt the statewide services to their own state by using the Internet or phone book to determine the phone numbers of the organizations within the state.

The final chapter in this section focuses on a specific environment —the college campus—and begins with a discussion of the evolution of campus policing. It advocates that contrary to popular belief, the college campus is not the safest environment. It too can have victims of crime. Bromley and Fisher provide a comprehensive discussion that entails media coverage, civil court decisions, federal and state legislation, crime statistics, survey results, and services available to college students, faculty, and staff. They complete their discussion by examining challenges to campus police that include making senior administrators aware that crime does occur on their campus, reducing barriers to victims' services, recognizing that "stalking" is a real threat, and ensuring that "underserved victims" receive recognition and equitable treatment.

Overall, this text has offered a variety of useful and important information about victims. Because police officers are among the very first "governmental" representatives to come in contact with victims, it is extremely necessary that they fully comprehend the importance of their interactions with victims, and that, at a minimum, they learn what services are available in their community and how to assist victims with obtaining such services. However, police officers are not the only individuals who must be well versed in dealing with victims. They are merely the first step in the healing process. Information about victims and how to assist them is important for all actors in the criminal justice system.

REFERENCES

CASA of Hidalgo County, Inc. (1999). *CASA: A powerful voice in a child's life.* Hidalgo, TX: Author.

Finn, P., & Lee, B.N. (1987). *Serving crime victims and witnesses.* Washington, DC: National Institute of Justice.

Greenfeld, L.A. (1996). *Child victimizers: Violent offenders and their victims—Executive summary.* Washington, DC: Bureau of Justice Statistics Clearinghouse. Available: www.ojp.usdoj.gov/bjs/pub/cvvoatvx.txt

Moriarty, L.J., & Kenworthy, P.C. (1998). Child representation models: A descriptive analysis of Virginia and North Carolina prototypes. In L.J. Moriarty & R.A. Jerin (Eds.), *Current issues in victimology research* (pp. 141–151). Durham, NC: Carolina Academic Press.

Pickler, N. (2000, March 1). Charges unlikely in first-grade shooting in Mich. *Chicago Tribune* [online]. Available: cnews.tribune.com/news...story/0,1235,tribune-nation- 54514,00.html